HIGHWAY OF DREAMS

A Critical View Along the Information Superhighway

TELECOMMUNICATIONS

A Series of Volumes Edited
by Christopher H. Sterling

HIGHWAY OF DREAMS

A Critical View Along the Information Superhighway

——— ◇ ———

A. Michael Noll

With a Foreword by
Abe M. Zarem

Routledge
Taylor & Francis Group
New York London

First published by Lawrence Erlbaum Associates, Inc., Publishers
10 Industrial Avenue
Mahwah, New Jersey 07430

Transferred to digital printing 2010 by Routledge

Routledge

270 Madison Avenue
New York, NY 10016

2 Park Square, Milton Park
Abingdon, Oxon OX14 4RN, UK

Library of Congress Cataloging-in-Publication Data

Noll, A. Michael.
    Highway of dreams : A critical view along the information
superhighway / by A. Michael Noll.
        p.   cm. -- (Telecommunications)
    Rev. ed. of: Highway of dreams : a critical appraisal of the
information superhighway. c1995.
    Includes bibliographical references and index.
    ISBN 0-8058-2557-6 (c : alk. paper). -- ISBN 0-8058-2558-4 (pbk. :
alk. paper)
    1. Telecommunications--United States. 2. Information superhighway-
-United States. I. Title. II. Series: Telecommunications (Mahwah,
N.J.)
HE7775.N65 1996
384--DC20   96-41620
                                                          CIP

Contents

i

FOREWORD

Abe M. Zarem

For the last several generations, we have been living through an extraordinary period of scientific and technological growth. In the last decade, economic, financial, social, and political alteration and expansion have occurred worldwide and at a pace heretofore unknown to us. As technologists, we observe that one of the most impressive areas in which dramatic changes have been occurring is in the field of communications and telecommunications. In fact, it appears that no area of technology is, at this time, changing more rapidly or more dramatically than the area of telecommunications. Evidence of this can be seen on all sides—from professional and public media attention, to speculative fever in the financial marketplace. It appears we are now living through an interval which I have dubbed "Tulip Time in Telecommunications."

For those not aware of it, the economic phenomenon known as Tulipomania exploded in Holland over 350 years ago. For a period of several years, an extraordinary fever of speculation drove tulip bidding and prices to a fever pitch while, in most cases, neither the bulbs nor the flowers were in evidence. During that interval, estates and fortunes were mortgaged with sales and resales of tulip bulbs being made over and over again, while the bulbs were presumably planted and neither bulbs nor flowers were to be seen. It was only natural that national scandal would result and economic disaster follow, with final intervention by the government. Enormous fortunes were lost.

Economic panics have a long history, and up to the 19th and 20th centuries, such financial disturbances and wild stock speculations were the result of, and were based upon, mainly shortages of goods and services and, frequently, of the unfounded anticipations of expanding markets, which it was hopefully thought would lead to great profits.

With the explosion of science and technology and the development of a financially strong middle class and the consequent emergence of mass market phenomena, as well as developed economic systems giving great rewards to entrepreneurs and successful risk-takers, a new phenomenon arose—the era of the new business startup and the "Business Plan." Driven by the more general availability of venture capital, larger, more fluid, and more flexible stock conditions, and armed with the figures showing revenue growth based on predicted sales, anyone with a simple computer could produce a financial spread sheet and Plan showing the anticipated business potential for profits, based upon assumed market size and future penetration for given products and services, the proprietary state of the basic technology to be employed, the status of competition, and so forth.

In the area of telecommunications, new business startups, alliances, acquisitions, and mergers blossomed everywhere. The words "digital electronics, advanced compression technology, multimedia, convergence" were projected by the media and became household words. So intense were colorful descriptions of new possible business opportunities that a new vocabulary was generated to cover the products, services, systems, and hardware that everyone was told would imminently arrive to affect our lives. Even children spoke enthusiastically of "cyberspace" and "hyperspace."

Frontier market experts emerged everywhere with predictions of future expanded consumer, business, or government needs—not always with justifiable or realistic estimates that could stand the test of careful inquiry or of time. Anticipated progress was based on market projections, the basis for which were never carefully investigated and,in many cases, were not appropriately related to historic or realistic purchasing patterns. So convenient and available were the new computer skills that business plans could be modified with the flip of a wrist, or the tap of a few fingers. The fine art of the developed Strategic Business Plan arose. Everyone seeking funds to create a new business entity knew they had to have such a plan. This was generally all to the good—for with this phase of activity came emphasis on the need for additional and new managerial and business skills.

So, in the last several decades, entrepreneurs and financiers who had witnessed the mighty march of miracles from our nation's laboratories, which, during World War II, were spurred on to develop new technologies that would help maintain us as a free people, now cried out for the peacetime technology transfer and the cost-effective creation of a new world of products and services for everyday consumer, business, and non-military use.

There have been many examples of the impact of scientific discovery and technical research and development as profound forces leading to economic growth, increases in our standard of living, and improvements in the quality of our lives. Powerful sources of new information, reservoirs of new knowledge, and new technologies have combined to provide the mainspring for new advances in energy, transportation, medicine and biology, and the production of food. But a reasonably good case can be made that nothing, at present, is more fundamentally important to our future than what we are beginning to see emerging in the field of telecommunications.

Our challenge is clear. How do we best apply ourselves to most effectively take the fruits of our intelligence and skills and make them come alive in a business sense to produce a cornucopia of telecommunications services and products which will cost effectively and successfully serve the expanded and main purposes for which they are intended?

How do we transition from the *Highway of Dreams* to *realistic dreams of the future* so that we may maximize what is going on today in telecommunications in order to improve appropriately the future of world society tomorrow?

Our principal need is for **strategic thinking**, conceptualization, innovation, the creation of new ideas, and the acceleration of additional creative processes. Such strategic thinking must be done in an environment of realism based on our knowledge of past experiences and in a multidimensional sense, including all of the principal factors upon which business success depends. This is particularly true for the present environment in telecommunications.

Specifically, we must have an understanding of what has gone before, in a business sense; the nature of human needs and wants; consumer psychology; commercial and business drivers and considerations that encourage success; government involvement in regulation and policy; finance; and, of course, technology. All of these must be considered together as a family of related factors, and they must be understood, if business success is to be assured.

The exposure and training of our youth to the necessity for emphasis on reality and understanding as well as strategic thinking is essential. The development of leaders who can conceptualize, innovate, and motivate completes the necessary combination for both program and business success. In the new era facing us, the role of the *leader* and the *strategic thinker* will be that of the effective *"mentor* "and *"tormentor"* so that the appropriate conversion of creative thought to realistic plans for action that will lead to business successes will emerge.

In essence, that is what this book by Michael Noll is directed toward.

Highway of Dreams fulfills an important and practical need. Its basic message is one of pragmatism and encouragement. Its message is that planning for the future must include practical reality based upon historic backgrounds and understanding. The status of technology and its evolution is only one factor in the development and acceptance of services and products in the field of telecommunications. The development of Strategic Business Plans must be preceded by strategic thinking of a basic sort which takes into account not only financial and technical considerations, but, also, the nature of customer needs and wants; government policy and regulatory conditions; manufacturing, sales, installations, and service costs. It must also consider experiences from the past with understanding of the causes for successes as well as for failures. This is done not to spread negativism or obstructionism, but, rather, to breathe reality and clear thinking into pulling together all of those conditions which go about to make business successes. This book will bring valuable historic perspective to its readers. It should cause discussion, debate, and creative thinking about the future in a broad sector of telecommunications. It is intended to challenge and motivate its readers to engage in strategic thinking about the "information superhighway" and deserves wide readership.

ABE M. ZAREM
Beverly Hills, California
June 1996

PREFACE

As I have become older, I have become more disturbed that so many people know so little about history, particularly when it comes to telecommunication. The danger in this lack of knowledge of history is that we are doomed to repeat the mistakes of the past. As an educator, I feel compelled to do something about such ignorance and hence have accepted invitations from various financial and business groups to give luncheon speeches that dispel the hype of the superhighway. As a way to reach a larger audience, I decided to use the overall structure of these speeches as a framework for a book that explores the reasons for my skepticism in much more detail. And hence this book was written.

The perspectives that I bring in writing this book span my entire career—a career that began at Bell Labs in the 1960s where I was doing research in speech processing and in the earliest use of computers in the arts. My research there lead to some patents, one of which covers the three dimensionality and force feedback of today's virtual reality. I was and still am a firm believer in technology and in the importance of new ideas.

My career took a major turn in the early 1970s when I left the basic research of Bell Labs and joined the staff of the White House's Science Advisor. Here I quickly learned that economics, policy, and societal forces are as important as technology in shaping the future. My colleagues at the Office of Management and Budget showed me how to ferret out the fat in an agency budget request. I also learned that many Federal bureaucrats are caring and competent people dedicated to the good of the Nation. I learned how people at the top were isolated from knowing what was really happening under them. I remember one day leaving my office at the Old Executive Office Building and walking out onto Pennsylvania Avenue realizing that the people standing in line for the bus probably know more about what was happening than I and had more valuable opinions than all the experts at the White House's disposal.

In 1973, I returned to Bell Labs and began research into the social factors affecting communication. The various policy issues involving telecommunication fascinated me, but my iconoclastic views that AT&T should divest manufacturing or risk losing the local phone companies got me into trouble with my management at Bell Labs. So I transferred to AT&T where I became involved with marketing and the planning and evaluation of trials for new services. One of my roles at AT&T was to analyze the trials and projects being performed in the department and suggest new approaches. At AT&T, I learned about consumers and management. But I also learned that trials can acquire a life of their own.

While at AT&T, I became involved with teaching the fundamentals of technology to my colleagues and also to students at NYU's Interactive Telecommunication Program. In 1984 I left AT&T and joined the faculty of the Annenberg School for Communication at the University of Southern California. I was dean of the School for a two-year interim term and learned about university administration and the academic battles that can occur over exaggerated nothings. I have also learned much about the need for vision of the future, guided by the realism of past history. And now as an old curmudgeon, I seem to live more and more in the past as I attempt to apply the past to understand the future.

Much of this book represents my personal opinions. However, these opinions are a result of personal observation and study, and as much as possible, I share with you the rationale behind them based on anecdotal evidence, historical research, and factual analysis. My hope is that my opinions will challenge you to question, rethink, and reform your own views.

This book is not an academic tome but is a work of practical knowledge and experience—it is a critical view of the information superhighway. However, I use this book as a basis for a graduate seminar course, augmented with current readings. The book can be used to present a balanced response to books that present Utopian views of telecommunication.

The book presents five factors that shape the future: technology, consumer need, finance, business, and policy. I am a firm believer that these five factors are the essential components, or framework, of a multidisciplinary approach to professional education for a career in communication and media management.

A small portion of the material in *Highway of Dreams* has been published in other places. Recently, I have been writing articles for *Telecommunications* magazine, where I have my own column entitled "TelePerspectives," and also for *The Sunday Star-Ledgerr* newspaper in New Jersey. Some of these articles are based on material and themes presented in *Highway of Dreams*. The section "The Extraterrestrials Are Coming: Leos, Geos, Meos" of Chapter 12 was previously published in *Telecommunications Policy*, Vol. 20, No. 2, 1996, pp. 79-82 and is reprinted with the acknowledgment of Elsevier Science Ltd., Kidlington, UK.

A number of people have in some way or another contributed to this book, although I alone take responsibility for its final content and conclusions. Or, as is frequently stated: "The views in this book are solely those of the author and do not necessarily represent ..." I appreciate the comments, suggestions, and encouragement of my colleagues, in particular, John Carey who suggested and contributed useful ways to present data on consumer spending and product diffusion, and Martin C. J. Elton, Eli Noam, Joseph J. Horzepa, and Mike Uretsky who all read various versions of the manuscript and made helpful comments.

The following people helped me during my background research for various portions of the book: Thomas Robak of Apollo CableVision; Edwin F. Comstock of Sammons Cable; Michael Luftman of Time-Warner; Philip Walker of Sprint; David Staudt of the National Science Foundation; Donald B. McCullough of Broadband Technologies; David Schoen of Orbital Sciences; Dennis Bone, James B. Donnelly, and Michael Losch of Bell Atlantic; Lisa Lapore of AT&T; Cree Edwards of CellNet; and many others whose omission should not be taken to mean that I do not appreciate their time and assistance. John Kollar and Leslie Sept at the Columbia Institute for Tele-Information helped me with some of the initial library research for the book. And lastly, I appreciate the encouragement, useful comments, and Foreword to this book of my mentor, tormentor, and friend Abe Zarem.

A. MICHAEL NOLL
Stirling, New Jersey
August 1996

Chapter 1

INTRODUCTION

The telecommunication and information superhighway is sweeping the country and the world. The superhighway is daily front-page news and promises a multimedia cornucopia of cyber and tele services that will revolutionize our lives at home and at work. Those who stand in its way will be run over by all the superhighway traffic from an integration of voice, data, image, and video services. Multi billion dollar deals are being formed as various companies in the communication industry form alliances, mergers, and acquisitions. We wonder whether all this really makes sense. Do we accelerate onto the superhighway or risk being left behind in a cloud of semiconductor dust? We are told: have faith in the superhighway; don't worry about its costs; build it and the users will all come; don't be a Luddite.

We wonder whether to view the superhighway with awe or with fear of yet another technology that we must learn all about. We vaguely remember the "revolutions" of only a few years ago, such as high definition television (HDTV) and the videophone, which have so far failed to materialize and wonder whether the superhighway—whatever it is—will suffer a similar fate. Is it sense or nonsense? Hope or hype? Reality or fantasy? What really is it?

Indeed, the definition of the superhighway is fuzzy, but it seems to refer to a telecommunication infrastructure that can deliver a wide variety of services, such as telemedicine, home banking, shopping, telecommuting, electronic mail, electronic data banks of information, video on demand, interactive entertainment, remote meter reading, and tele-education. The superhighway is characterized sometimes as a broadband interactive network—whatever that might be. Another characterization is as a single broadband medium to the home, most likely consisting of optical fiber. Yet another characterization is as the convergence of entertainment video with interactive, two-way telecommunication and the convergence of video entertainment with computers. It is the Internet and the ability to gain access to all the world's information.

Our minds begin to swirl with technological concepts that seem beyond comprehension. If it is high-tech, then it must be wonderful, we are told.

If you wonder what the superhighway really is; if you question whether there is really anything new or wondrous about it; if you have doubts about accelerating onto an unknown superhighway and about where it might take you; then this book is for you.

We explore the history of the superhighway, going back to its roots in the late 1960s and early 1970s. We question whether people really want all its marvelous, supposedly-new services. We learn about the basic concepts of modern communication technology and whether all that is promised by the superhighway is really feasible technically. We explore its costs; whether there are any real profits in it; who might be able to make it happen; and what policy and regulatory barriers stand in the way.

So that the suspense does not overwhelm you, allow me to disclose my conclusion that the superhighway is a lot of hype and fantasy, promising services that most people do not want, nor are willing to pay for; that the superhighway would be costly to build; that much of the technology exists only on paper and is not real; that its construction might result in a total monopoly of entertainment and telecommunication by a few, super-large firms. Let me also share a secret with you: in many ways the superhighway is already here! It is the network of networks composed of today's switched public telephone network, industrial private networks, public and private packet-switched data networks, CATV cable to our homes, and communication satellite transmission of TV programs.

As you can well imagine, my critics accuse me of being too negative; of being a Luddite; of having no vision or faith. To them, I say faith belongs in church. I warn them that their Utopian vision is old hat and will for some become a financial nightmare.

Risks must be taken to explore and develop new products and services. While I was at Bell Labs, my management taught me the importance of always having the freedom to fail—the freedom to explore new areas even when the path might ultimately result in a dead end. However, much of today's superhighway is not new but is a resurrection of past failures, perhaps in ignorance or perhaps in the blind faith that the times have changed and are now ripe for success. All the hype and hysteria along the information superhighway have created paranoia, fear, and panic to such an extent that any real opportunities will surely be missed. A fog is settling over the superhighway, and a big crash is about to occur as all the greedy participants rush blindly along until they finally rear-end each other.

But hype and promotion are nothing new. Every new product or service was promoted vigorously by its advocates. Computer firms, Wall Street, and Hollywood—to mention only a few—are all guilty of hype and over-promotion. Even Thomas Alva Edison continued to promote his direct current method of electricity long after alternating current was proven superior.

Just because some product or service has failed in the past is no reason in and of itself that the failure will continue forever. A good example of this is facsimile. The basic idea for fax is older than the telephone, but fax only achieved its recent market success during the past ten years or so. However, this was because the fax machines of the past suffered from serious technological and cost deficiencies. Once these deficiencies were solved fax finally took off after a 100-year history of relative failure. So, history alone does not shape the future; other factors are involved too.

Although much of this book might seem to be about the information superhighway and telecommunication, it is also about understanding the future. I suggest two ways of obtaining such an understanding. One way is to study and understand the past. History has a way of repeating itself. Although we should not be trapped in history, ignorance of past mistakes indeed dooms us to repeating them.

The second way to understand and evaluate the future is to perform a reality check by analyzing the five factors that I believe truly shape the future—namely, technology, consumer needs, finance, business, and policy. The overall reality of any business venture or some new product or service is only as strong as the weakest link created from these five factors. I hope that the use of these five factors will help you learn a broad way to analyze new ventures beyond just their use in analyzing today's superhighway.

So that we do not end the book with too much of a negative taste, I conclude with a discussion of the exciting telecommunication services that are already here today and that we take much for granted. In this respect the superhighway becomes simply a new term for what already exists.

We begin this book by painting a picture of the technological telecommunication Utopia promised by the superhighway. Hold onto your hats, this will be a fast and exciting cyber-drive at warp speed down the superhighway as the electronic billboards zoom by: ISDN, HDTV, virtual reality, video on demand, cyber-space!

READINGS

Connell, Stephen, "Broadband Services in Europe,"*Telecommunications in Transition*, eds. Steinfield, C., J. M. Bauer, and L. Caby, Sage: Thousand Oaks, CA, 1994, pp. 236-251.

Hayashi, Koichiro, "Information infrastructure: Who builds broadband networks?" *Information Economics and Policy*, December 1993, pp. 295-309.

McCarroll, Thomas, "A Giant Tug-of-Wire," *Time*, February 24, 1992, pp. 36-37, 41.

Noll, A. Michael, "The Broadbandwagon," *Telecommunications Policy*, September 1989, pp. 197-201.

Phillips, Barry W., "Broadband in the Local Loop," *Telecommunications*, November 1994, pp. 37-38, 40, 42.

Chapter 2

A TECHNOLOGICAL UTOPIA

It is much fun to peer into the crystal ball to see a technological Utopia facilitated by modern electronics and telecommunication. This is a Utopia of many wonders under the assumption that technology can accomplish almost anything and that nothing. This chapter presents that Utopia with its wondrous world of new services and products. Join me in letting our imaginations run free as we race along the superhighway. Some of what we imagine about this technological Utopia is real but much is fantasy.

AT HOME

The superhighway will have considerable impact on entertainment and other activities at home. Television will be transformed into a true multi-media extravaganza of home entertainment with high-quality digital surround sound and super-size high-definition images. The television image will fill a whole wall of the home with a thin panel capable of showing high-definition images, either throughout the whole display area or by tracking the eye movement of the viewers and focusing maximum resolution on the portion of the image being viewed with less resolution for those portions of the image in the visual periphery. Personal television will be accomplished with television sets the size of a book and with high-definition images on thin-panel displays.

Video entertainment will be delivered to the home over a wide variety of media, some electronic and some physical. Rather than being presented with only 50 channels of cable television programs, viewers will be able to choose whatever they wish to watch whenever they wish to watch it—a system known as video on demand. Central storage of movies and other forms of video programs will be accessed and transmitted individually to each viewer's home on demand—a form of video dial tone. All the movies ever produced will be available in this fashion with no need to venture from the home to a video store.

Today's video store—like the telegraph office—will be a relic of the past. We will "dial up" TV programs and movies from around the world, and these programs will be switched to us over a broadband network which will offer such flexibility and programming options that conventional cable television will be unable to compete and will disappear. Conventional broadcast over-the-air television will likewise disappear as TV signals are delivered on-demand to us over broadband optical fiber.

Hundreds of channels of television programming from around the world will be sent to our homes from communication satellites located in geosynchronous orbits 22,300 miles above the earth's equator. A small antenna at our home will be electronically steerable to aim itself to receive the signals from each satellite. The CD-quality audio will be sent in many languages so that you could choose your language of choice. Imagine being able to watch soccer from Italy, Wagner from Berlin, a game show from Spain, or news from the BBC at any time in your home and in your language of choice.

The world's weather patterns over the past 24 hours will be instantly accessible so you can predict your own weather. Traffic patterns will be transmitted to your home before you depart for work. Global positioning will guide you in your automobile, and paper maps will become museum relics.

The family will sit in front of the large screen and their image will be sent to distant locations while they simultaneously see the other people—a form of video teleconferencing. Grandparents will be able to see, talk, and visit with grandchildren without ever leaving the home.

Rather than watch the TV camera image decided upon by the TV show's director, you will be able to choose your own camera shot while watching sporting events, talk shows, and other TV shows. If you want to focus on a guest on a talk show, you will be able to do so. If you want to watch only your favorite sports hero during a football game, you will be able to do so. If you discover on the subway that you forgot to set the VCR, you will be able to program your VCR using telephone and wireless telecommunication facilities.

CD-quality audio will be stored on three-dimensional digital memories that plug into a player without any moving parts that could wear out or require maintenance.

Broadcast radio will be replaced by direct radio in which the listener "dials up" the requested programming. All the world's recorded music will be instantly available on demand from central storage banks. There no longer will be a need to visit a music store to purchase a CD or tape. Private libraries of recorded music, like private libraries of VCR tapes, will disappear.

All the world's knowledge will be instantly accessible from computer terminals over a packet-switched public data network that is friendly and easy to use. Students will do their library research without ever going to a library. Books will all be stored at central data banks in a digital format and will be accessible over the telecommunication superhighway. The paper-based book will disappear and will be replaced by a thin-panel electronic display that will have the look and readability of paper.

Travel reservations will be made directly by you, eliminating travel agents and airlines reservation clerks. Anyone planning a vacation or trip will be able to tele-visit their destination in advance by navigating the streets of distant cities or hiking the trails of Yosemite through the technology of virtual reality accessed over the superhighway. Zoos will be able to concentrate on preserving endangered species since many people will tele-visit the zoo.

Shopping will be revolutionized by the superhighway. One will walk along an electronic mall and will tele-enter various stores and shops to see the various goods and items available. Force-feedback gloves will allow you not only to see but to pick-up and feel goods and clothes. Computer simulation will place you in a new wardrobe and allow you to see yourself in an electronic mirror along the superhighway. Redecorating will be a dream of comparing many options in a computer-simulated rendition of your bedroom before a can of paint is ever applied.

Orders for goods and food will be made electronically from the home without ever visiting a store. Comparison shopping for features and prices will be a breeze involving no more effort than entering a few keys at the home terminal. Your body will be scanned by a camera at your home terminal and your clothing size automatically determined.

You will then be shown computer-drawn renditions of yourself wearing any desired clothing before you decide to place your order. Looking for a new home will be a breeze along the superhighway as you search computerized files for the home matched to your specific lifestyle profile and then look at videos of the homes that are your perfect match.

Newspapers too will be revolutionized, and fingers will no longer be black from ink since paper and ink will disappear. Electronic information will be accessed and transmitted over the superhighway. An electronic profile of your individual interests will be gradually created and only those news items and stories of particular interest to you will be sent to you each day. The news in text and images, even moving images, will be sent during evening hours and available in the morning for display on a thin-panel terminal. These terminals will be portable so they can be carried and read by those few people who still commute physically to work.

Paper money will likewise disappear, and all transfers of funds will be transmitted electronically over the superhighway. We will carry electronic money cards that can even be accessed automatically, for example when passing though an electronic turnstile in the subway without the need to even take the card from your pocket or purse. The funds in the card will be renewed over the superhighway by inserting the card in a home terminal and then obtaining the renewal of the funds by entering appropriate key strokes and authorization codes.

Junk mail will be a distant memory of a past world of physical delivery. All mail will be electronic over the superhighway. Only those advertisements and announcements that one wishes to receive will be stored and saved. A junk-mail violation will result in instant electronic prohibition for all time and messages from the offender will never ever again be received.

The trip to the polls to vote will be replaced by instant electronic voting over the superhighway. Opinions of people will be gathered constantly on all sorts of political and public issues—a new electronic town hall and a form of teledemocracy. Lottery tickets will be purchased over the superhighway, and tele-betting will be conducted at home on a wide variety of sporting events.

Much physical delivery of information will be replaced with electronic delivery to your home over the superhighway. Games will be downloaded to your home game unit over the superhighway. You will be able to play games with other, distant people, not only sharing the same game but also seeing and hearing your opponent. You will choose opponents based on your skill level, perhaps with appropriate handicaps.

Magazines, books, and many of today's paper-based media will be replaced with electronic delivery over the superhighway. Software will be downloaded over the superhighway. The amount of such data traffic will greatly exceed the voice traffic carried over today's telephone network.

The electronics in the home will be interconnected over a home broadband network and will be programmable at a central computer in the home. Programming of the computer will be accomplished by simple voice commands which will be understood by the computer. Voice dialing of the telephone, audio announcements of the name of the calling party, and audio responses to verbal instructions will make the user interface natural, friendly, and easy to use. A phone call to a foreign land will be as easy as calling next door since foreign speech will be automatically translated into your favored language, and vice versa.

The home telephone will combine facsimile and data communication with voice communication for transmission over a digital communication highway. A display panel on the phone will announce the name and address of all callers, show electronic messages and bank balances, and assist in the programming of intelligent-network features.

Gas, water, and electric meters will be read electronically so that you will no longer have to be home to give access to a human meter reader. This electronic remote meter reading will occur every minute or so thereby allowing the utility company to offer different rates depending upon demand during the day. The home control computer will access this meter information and will adjust the use of water, electric, and gas accordingly to optimize price with usage needs.

Electronic monitors in the home will protect from flooding, fire, and unauthorized entry. The state of the home will be accessible over the superhighway over distance so that you will know your home is secure even when you are not there.

Graphics-based virtual reality will facilitate your searches for information along the superhighway and your selections of entertainment. The dial and remote of today's TV set will disappear. Choices will be determined automatically by where you are looking, perhaps by reflecting a laser beam off your eye. Ultimately, brain waves will be analyzed and choices made simply by thinking about what you want to watch or listen to.

AT SCHOOL AND STUDY

Education will be changed greatly by the superhighway. College and university students will be able to dial up lectures around the world at the world's greatest universities and centers of knowledge. These lectures will be delivered in video over the superhighway. The same video technology will enable high school and secondary school students to attend courses by video and distant teachers to see their video students over two-way video links.

Many students will study at home and dial into the world's electronic libraries of information. Text, graphics, pictures, and even moving images will be accessed and sent over the superhighway to the students in their homes and dorm rooms. When additional instruction is desired, students will dial up video lectures by the world's experts and these lectures will be customized to the student's level of understanding. The educational experience will be greatly enriched.

Students working together on projects will send electronic mail, or e-mail, messages to each other and to their instructors thereby facilitating team work that transcends time and distance. The video classroom will extend to home television sets that connect to the superhighway, and students will dial up specialized educational courses to create an open global university of learning without the need to travel to a specific school or university.

AT WORK

Many people will work at home and will keep in contact with customers and colleagues over the superhighway. E-mail messages and two-way video telephone calls will keep workers in contact with each other and their customers, creating closer and more frequent relationships than were ever possible with physical presence. Groups working on projects will meet by two-way video teleconferences without the need to travel. Participants will be able to look wherever they wish, in effect being teleported through telecommunication to the distant meeting room.

The office will be transformed into an array of electronic displays and computer equipment. All the information needed to do your job will be available instantly over the superhighway, while at the same time you are in instant contact with your customers and colleagues. The city will no longer be a center of business since rural and suburban centers will be interconnected over the superhighway. A screen of multimedia information will present graphs, charts, and video images of the people being reached and the information being exchanged.

Some labor-intensive service work will be exported to foreign countries. One such example is telephone based directory assistance. Calls to directory assistance will be forwarded to operators in foreign countries that have low labor costs. This will be possible because of the low transmission costs of telecommunication. Data entry work is already being done in foreign countries with low labor costs, and low-cost telecommunication will simply extend this practice to other services. Ultimately, automatic speech recognition will automate even these tasks so that human intervention will no longer be needed to obtain a telephone number or perform a routine transaction.

EVERYWHERE

The intelligence of the computers that control the superhighway will know where you are at all times. Thus, any phone call will be forwarded to you automatically wherever you are at any time of the day or night. You will program this access so that different callers will have different levels of access to reach you. When you do not wish to be reached, voice mail will intercept the caller so that a message can be left for you.

Banking will be transformed by the superhighway. No longer will you need even a dime. Electronic money will be as easy as inserting a plastic card into a slot. The bank will be available over the superhighway from home, work, and everywhere. Funds will be transferred, balances determined, loans granted, and investments made without ever physically visiting a bank or investment office. The superhighway will bring banking to you everywhere, just by entering a few keystrokes. Cyber cash will be secure and protected by unbreakable encryption codes.

AT PLAY

Virtual reality with sophisticated simulations involving graphics and motion will bring a revolution to play. You will fly a Boeing 747, feeling the bumps as you land. You will fly supersonic fighters in simulated battles, being rattled by near misses from missiles. You will swing the club on the most challenging golf courses ever imagined. You will stalk and attack opponents in simulated battle.

Movies will become multimedia extravaganzas. Three-dimensional images, smell, and feel will bring you into the action. Super-definition movies in stereoscopic imagery will be transmitted to the theaters over the superhighway rather than in physical reels of film. Film will disappear as a medium. Cameras will store images electronically for display on a TV screen or small, hand-held, pocket viewer. Virtual reality will introduce fantastic pornographic experiences—a form of cyber-porn.

ON THE ROAD

Car theft will never occur again. Your car will recognize only the speech of authorized drivers and will not start until the identity of the driver has been confirmed through speaker verification. The car will react to verbal commands, such as "ignition," "wipers," and "lights." You will never be lost on a freeway. Navigation will be based on precise positioning data from geosynchronous satellites, and detailed electronic maps will be displayed on the windshield superimposed on your view of the road. You will be guided onto appropriate exit ramps, around traffic congestion, and merged into the appropriate lane. TV images of key traffic locations will be sent to your car so that you can determine the best route to your destination.

On long trips, your car will be automatically driven along electronic guides imbedded in the roadway. Steering, acceleration, and braking will all be automatic. You will simply sit in the your vehicle and enjoy the computer-controlled drive.

WHEN ILL

Visits to your physician will be conducted over the superhighway. Your physician will be able to see you on your videophone, and sensors will measure such things as temperature and blood pressure, thereby eliminating much of the need for a physical visit. Your physician will exchange information with the medical experts of the world, performing consultations remotely over the superhighway. X-rays, CAT scans, lab reports, and other data will all be digitized for instant transmission from one specialist to another. All the world's medical expertise will be available wherever you are, from the largest cities to the most remote rural community.

Micro-miniaturized "machines" will enter your body to seek out clogged arteries and to eat away obstructions. Other such machines will perform surgery automatically within your body, guided by computers and the hands of a surgeon assisted by a virtual reality depiction of your innards. All your medical data and information will be encoded on a chip on a card that you carry in your wallet or purse. The information will be instantly accessible in any medical situation or emergency.

AT THE ARTS

The widespread availability of personal computers with software to compose music, to choreograph ballets, to animate movies, and to paint pictures will result in a new form of citizen artist who will be able to distribute art over the superhighway to the entire planet as an audience. Novels will be written and distributed instantly over the superhighway eliminating the roles of the publisher and book store.

You will be able to tele-visit the Metropolitan Museum of Art by electronically walking its galleries and seeing the various works of art, along with hearing descriptions of the works being seen and then detouring to learn about the life of the artist. When you rarely venture from the home to attend an opera or play, the stage scenery will be computer-generated, three-dimensional holograms.

AND NOW FOR SOME SOBERING HISTORY

This then has been a short but high-speed trip along the superhighway's dream of a technological Utopia of a cornucopia of telecommunication services. It all sounds great, but if you are as old as I, much of it sounds all too familiar. You wonder whether people will really want all these services, whether they will afford them, whether industry will really be able to deliver them or afford the required investment, and whether governmental policies will allow them.

The next chapter reviews history and shows how this vision of a technological Utopia is a resurrection of many old ideas that have consistently failed to achieve the revolutions promised by their promoters and advocates in the past.

READINGS

Adam, John A., "Special Report: Interactive Multimedia," *IEEE Spectrum*, March 1993, pp. 22-31.

Bernier, Paula, "FCC OKs Nynex, BellSouth video dial tone applications," *Telephony*, February 13, 1995, p. 6.

Britt, Russ, "Pac Bell to offer vast data service," *Los Angeles Daily News*, November 12, 1993, pp. 1, back.

Brunet, Craig J., "Viewpoint: Electric utilities can jump-start the Infobahn," *IEEE Spectrum*, January 1995, p. 28.

Comerford, Richard & Tekle S. Perry, "Wired for Interactivity," *IEEE Spectrum*, April 1996, pp. 21-32.

Fenichell, Stephen, "NYNEX Marks the Spot," *New York*, August 29, 1994, pp. 24, 27.

Gates, Bill with Nathan Myhrvold & Daniel Paisner, *The Road Ahead*, Viking: New York, 1995.

Gilder, George, "The death of telephony," *The Economist*, September 11, 1993, pp. 75-78.

Martin, Robert L. and George C. Via, "Operations systems for the revolution," *Exchange* (published by Bellcore), December 1993, pp. 13-17.

Negroponte, Nicholas, *Being Digital*, Vintage Books: New York, 1995.

Newton, Harry, "The Information Superhighway," *Teleconnect*, May 1994, pp. 10-17.

Niggli, Michael R. and Walter W. Nixon, III, "Why Electric Utilities Should Build the Information Superhighway," *Telecommunications*, March 1995, pp. 68, 73-74.

Noll, A. Michael, "The Digital Computer as a Creative Medium," *IEEE Spectrum*, Vol. 4, No. 10, (October 1967), pp. 89-95.

Noll, A. Michael, "Man-Machine Tactile Communication," *SID Journal* (The Official Journal of the Society for Information Display), Vol. 1, No. 2, (July/August 1972), pp. 5-11.

Noll, A. Michael, "Teleportation Through Communication," *IEEE Transactions on Systems, Man and Cybernetics* Vol. SMC-6, No. 11, (November 1976), pp. 753-756.

Patterson, Jane Smith and Scott Anderson, "Take a Ride on Carolina's Information Highway," *Telephone Engineer & Management*, November 15, 1993, pp. 32-34.

Rebello, Kathy and Richard Brandt, "Your Digital Future," *Business Week*, September 7, 1992, pp. 56-61.

Rebello, Kathy and Robert D. Hof, "Interactive TV: Not Ready for Prime Time," *Business Week*, March 14, 1994, pp. 30-31.

Yu, Deborah, "Ventura's video future from Cerritos," *Los Angeles Daily News*, May 29,1994, Business Section, pp. 1,3.

Message from the Chairman, Bell Atlantic Shareowner Report, 1993, Third Quarter.

"Roll over Gutenberg," *The Economist*, October 16, 1993, pp. 105-106.

"So much for the cashless society," *The Economist*, November 26, 1994, pp. 21-23.

"Television: What if they're right?" *The Economist*, February 12, 1994, Survey, pp. 3-18.

"The interactive bazaar opens," *The Economist*, August 20, 1994, pp. 49-51.

"The third wire," *The Economist*, January 28, 1995, pp. 62-63.

Chapter 3

HISTORICAL PERSPECTIVE: LESSONS FROM THE PAST

An important way of understanding the future is to have a firm knowledge and grasp of the past. Very few things are truly novel, and many things are simply reincarnations of the past. As has been observed: "The past is prologue to the future." Those who forget the past are doomed to repeating its mistakes. Innovation means not repeating the mistakes of the past.

It is therefore important to know what has been done in the past regarding the many services and technologies of today's superhighway. We shall review this history in this chapter. We shall see that much of what is proposed today is not new and has been investigated quite thoroughly in the past. Not even the term "communications highway" is new—in fact, a 1909 AT&T advertisement described the Bell System as a "highway of communication."

THE BEGINNINGS OF TODAY'S SUPERHIGHWAY

Much of the rhetoric about today's superhighway involves entertainment and television, specifically television delivered over wires to our homes. In this respect, today's superhighway had its beginnings in cable television. Cable TV actually is almost as old as broadcast television and began as a means to obtain TV in areas not reached by the radio signals of broadcast TV. The radio signals were picked up by a large community antenna and then sent along coaxial cable to homes—thus the term CATV originally meant Community Antenna TV.

In the late 1960s, cable television, or CATV, was still in its infancy but was poised for explosive growth. Visionaries of the day foresaw opportunities for CATV to solve many social problems of cities through broadband communication.

In 1968, the New York City Advisory Task Force on CATV and Telecommunications, chaired by Fred W. Friendly, issued its report urging New York City to move forward in developing a city-wide complex of CATV systems offering two-way capability to provide a host of telecommunication services. That same year, the President's Task Force on Communications Policy, chaired by Eugene V. Rostow, issued its report urging greater use of CATV as a means to greater diversity of TV programming. These two reports became the impetus for policies at the local and national levels to speed the installation of CATV systems around the nation.

The report of the Sloan Commission on Cable Communications appeared in 1971 and promised a "television of abundance" that would come from wired broadband communication systems as CATV spread across the United States. In 1972, Ralph Lee Smith's book *The Wired Nation,* based on his 1970 article in *The Nation,* described all the wonders that would come from the continued development of cable TV and broadband communication. Smith claimed that cable TV was about "to effect a revolution in communication" by offering many channels of viewing, local origination, community broadcasting on open public channels, and ultimately a "single, unified system of electronic communication." He strongly encouraged community activists and civic groups to take an active role in the franchise process to promote his ideas for a wired nation and an "electronic highway."

In 1971, the National Academy of Engineering's Committee on Telecommunications issued the report of its Panel on Urban Communications, chaired by Dr. Peter C. Goldmark, then president of CBS Laboratories and, among other things, inventor of the long-playing phonograph record. This report, entitled "Communications Technology for Urban Development," was the result of a study by the panel of the possibilities for using telecommunication to improve urban living and stimulate urban development. The report saw telecommunication as a solution to many of the problems of cities in such areas as medical care, education, public transportation, and law enforcement. It suggested that in addition to the telephone network, the CATV network could be used also for sending information to homes and for a return data capability.

The report suggested "a broadband communications highway" interconnecting large institutions in the city. The report further suggested the conduct of pilot projects in a number of social areas.

The Goldmark study and report took a very positive view toward its concept of "broadband communication networks" created by adding two-way capability to a CATV system along with a centrally located digital computer. Home shopping, polling, emergency services, education with student response, home appliance control, and automatic meter reading were only some of the many services that were envisioned in Goldmark's telecommunication Utopia.

One area in which the Goldmark study and report were very valuable was in encouraging the investigation and the ultimate development of 911 service with a capability for automatically identifying not only the caller but also the caller's location (known as Automatic Location Identification, or ALI for short). In 1972, Goldmark proposed the use of a variety of communication technology, including broadband CATV, conventional telephony, and communication satellites, for serving rural areas of the United States in a presentation entitled "The New Rural Society."

As a result of all this continuing hoopla over CATV, many communities stipulated that the company proposing to provide CATV service would have to provide community access and two-way capabilities to obtain a franchise. The CATV companies promised much to obtain the franchise but then provided as little as possible once they had the franchise. But, as we shall see, the consumers could have cared less, since there was very little interest in any of the possible two-way services nor was there much interest in community access.

INTERACTIVE CABLE

In response to the need for public policy to promote the new uses for cable television, the Federal Communications Commission set rules in the early 1970s requiring that CATV systems of specified sizes be capable of two-way interactivity. In 1974, the National Science Foundation initiated a series of research projects to investigate the use of interactive television in providing social services. The major conclusion from the NSF projects was that the technology worked and was as effective as other methods. In addition to the NSF-supported research, private industry was also conducting trials of new interactive uses of CATV, most notably Warner's QUBE system, which will be described separately in the next section.

One NSF supported project involved three neighborhood community centers which were linked by two-way video. The system was installed in Reading, Pennsylvania to facilitate the dissemination of information to senior citizens. The project was conducted by researchers from New York University in cooperation with the local CATV firm, the city, and the local Senior Citizens Council. Local viewers called in by telephone, and the live discussions between the three locations were presented in switched-camera shots and split-screen fashion. At the end of the research project a non-profit corporation was formed to continue the operation of the system. It is still operating today, thereby testifying to its success. It was the first application of a collection of remote TV sites coupled with telephone calling from viewers—an early form of "talk TV."

The use of data return over CATV in response to questions was investigated for use in day-care training, high-school equivalency, and a child development course in Spartanburg, South Carolina. The project was conducted by the RAND Corporation in cooperation with the local CATV firm and the Spartanburg Technical College. The remote students pushed buttons on a home terminal in response to questions transmitted over the cable along with the educational programming. The two-way interactive communication was found to be as effective as conventional classes and one-way broadcasting. However, a greater number of students responded in the interactive situation. Also, travel to attend classes was avoided.

Two separate projects were conducted in Rockford, Illinois. One, conducted by researchers from Michigan State University, involved fire fighter education over a CATV system. Videotaped programs were transmitted over the CATV system with an interactive data return of typed responses on a terminal. This condition was compared with two more conventional alternatives: one-way video and video cassettes, both with mail response to the questions. In all three conditions, questions were asked of the students after a key point had been made in the training presentation. The interactive CATV approach was better in terms of test scores and retention after six months.

The second Rockford project, conducted by researchers from the University of Michigan, involved in-service training for public school teachers over interactive CATV with a data return, compared to traditional CATV. Standard classroom training was used as a control condition. Although the interactive CATV scored better than traditional one-way CATV, the standard classroom setting scored higher than the one-way CATV. The results of both Rockford projects were mixed, and the differences between the scores for different conditions were sometimes small.

The MITRE Corporation's TICCIT (Timeshared Interactive Computer Controlled Information Television) system began operation in Reston, Virginia in 1971. The telephone was used to request information from a computerized data base, and the requested information, such as a bus schedule, was sent over the CATV system for display on the home TV receiver. TICCIT was terminated in 1973. In the same time frame, nearly 200 homes in Woodlands, Texas were supplied by TOCOM with two-way cable for fire and alarm services.

WARNER'S QUBE SYSTEM

In late 1977, Warner Cable, a subsidiary of Warner Communications, Inc., began a test market in Columbus, Ohio of a two-way, interactive CATV system. Columbus was chosen because of its image and history as a progressive city. Columbus was frequently used as a test market for all sorts of new products and services.

QUBE offered 30 channels of TV programming, including 20 channels of "free" programs. The up-stream return signal was used for low-bandwidth data such as program selection for the 10 channels of pay-TV, game-show participation, balloting, and auction bidding. The monthly rate was $10.95 plus pay-TV usage. In 1979, QUBE was available to 38,000 subscribers of the 45,000 total subscribers in Columbus. QUBE was developed by Warner in collaboration with Pioneer Electronic Corporation which made the interactive terminal. Emergency and security services were planned for the future.

QUBE was very valuable as a marketing tool since it allowed Warner to poll the set-top box every 6 seconds to determine what channel was being watched. In this way, new programming could be tested and viewer response obtained almost immediately. The use of QUBE as a programming laboratory resulted in today's cable programming networks "MTV" (rock videos) and "Nickelodeon" (children's programs).

The talk-back capability of QUBE was not compelling to most viewers, and the impulse, pay-per-view feature was not a powerful revenue generator. Newer technology had been developed for CATV distribution, and QUBE became outdated. Thus, QUBE was costly for the revenue that it generated, and the interactive, talk-back feature was not that important to most viewers. Such services as electronic banking, shopping, and mail had little appeal to most people using QUBE. The end result was that QUBE service was terminated in Columbus in the late 1980's, and the last QUBE system, which was in Cincinnati, was terminated in 1994.

TELECONFERENCING

In the mid 1970s, at the height of the energy crisis, AT&T installed public-room, two-way, video teleconferencing facilities in a number of major cities in the United States. The rooms were well designed and included extensive video graphics capabilities. But even after an intense marketing effort, usage remained disappointingly low. I recall a focus group meeting in which an AT&T employee in New York stated that it was easier to take the train to the District of Columbia rather than to walk across the street to use the video teleconferencing system. In a later chapter on the pitfalls of trials, I describe how AT&T was unable even to give away the service to most potential users.

The experience at AT&T was typical. The Confravision public-room video teleconferencing system operated by the British Post Office in the United Kingdom and a system operated by Bell Canada had similar poor market acceptance.

What was ultimately discovered was that teleconferencing seemed best suited to a certain type of meeting. This type of meeting was recurring with participants who knew each other. The purpose of the meeting was mostly the exchange of information as opposed to the conduct of bargaining and persuasion. This suggested a target market for teleconferencing. I was able to estimate this target market as being about four percent of all meetings. Furthermore, two-way video teleconferencing was only one of a range of teleconferencing technologies that also included two-way speakerphones, permanently installed two-way audio, and two-way graphics.

This is not to say that there is no market for teleconferencing, but the market is smaller than the 20 percent to 80 percent of all meetings that promoters have predicted. My belief is that improved audio-only technology (such as hi-fi and stereo) augmented with interactive, two-way graphics could satisfy much of the market for teleconferencing without the need for two-way, full-motion video.

Audio-only teleconferencing conducted from a speakerphone or a special audio teleconferencing unit continues to be successful. Today, video compression has greatly reduced the bandwidth needed for the two-way video, which has resulted in greater use of on-premises video teleconferencing. Software is available for most personal computers along with small cameras to facilitate such desk-top video teleconferencing. However, the resolution is such that a viewer is just barely able to determine whether the eyes of the distant participants are open or closed. The video channel really adds little to the communication other then improving the ability to identify who is still in the meeting room or who is speaking.

The AT&T video teleconferencing system taught me the dangers of pricing a service far below cost. The AT&T video teleconferencing system of the 1970s offered full video bandwidth, and the service was offered at special low rates on a trial basis. A large bank in California expressed interest in the service for use between its offices in San Francisco and Los Angeles. The bank was particularly interested in using the service in the late evening hours. Since most meetings occur during normal business hours, we were puzzled by the interest of the bank to use the service at such unusual hours. What we discovered was that the bank intended to use the bandwidth not for video but to transmit large volumes of data between the bank's computers each evening. The video-teleconferencing bandwidth was much less expensive than data-communication bandwidth at the tariffed rates since the teleconferencing system was being offered at special trial rates that were less than actual cost. We had inadvertently created an opportunity for competition with Pacific Bell.

TELE-EDUCATION

One of the earliest instructional television systems was initiated in the late 1960s by Stanford University for broadcasting engineering courses around the San Francisco Bay area. The system broadcast one-way TV signals to a number of remote sites. The radio technology was known as Instructional Television Fixed Station (ITFS) and used the 2.5 GHz band that was authorized by the FCC for educational institutions and organizations. Stanford constructed special studio classrooms for the system. FM radio was used for talk-back to the instructor. The Stanford

Instructional Television Network (SITN) today has ten studios on campus and sends its courses by live broadcast, two-way video and also by tutored videotape.

The use of video for tele-education has always been quite successful but most certainly is nothing new. In 1975, about 60 ITFS systems were broadcasting a total of nearly 250,000 hours of instructional programming to public school and college students, medical professionals, and business people, according to data issued by the National Association of Educational Broadcasters. Most of the video systems are point-to-point and are what has been called closed circuit TV, or CCTV. The systems require much bandwidth and are not switched.

The role for a common-carrier telephone company in educational technology, other than providing transmission bandwidth, has always been questionable. In fact, almost any technology can be used for education. The key questions focus on cost. Few educational institutions would be able to afford switched access of mega-Hertz or mega-bits of capacity, nor would their point-to-point applications require sophisticated, costly switching.

Most educators realize that the personal interactions beyond the classroom are a key factor in education, particularly at the elementary and secondary levels. The use of video can be somewhat dehumanizing (perhaps explaining the accepted use of video in engineering, since so very few human dimensions are involved). A good textbook, an interactive computer program, or an audio tape can all be equally effective in education. The key factor is student motivation.

Few students paying $2,000 for a graduate-level course at a major university would feel they had received good value for their money if they sat at home and attended class by video. Learning involves interactions with other people, with other students, and with faculty. Education is learning how to think, how to analyze various problems, and how to interact with others.

CATV has been used for tele-education. As a franchise condition, some CATV firms provided free or at-cost service to local schools in the early 1970s. Two-way capabilities were tested to enable home-bound children to tele-commute to class to see and hear classes. Indeed, the technology worked, but in the end the real issues always involved cost, and still do.

Indeed, video can be used to deliver education, but so can a book, a radio program, an audio tape, or a computer. Whatever the medium, if the students are motivated they will learn. The specific medium, however, has no learning benefits, as shown in a very thorough survey paper by Prof. Richard E. Clark.

Tele-education is not a total failure, but it also is not the large market that some of its promoters had hoped. It, like video teleconferencing, has its place in a spectrum of delivery technologies. Video tele-education is expensive compared to other technologies and thus continues to be limited to high-need applications, like graduate engineering education to remote sites. Technology has its place in the classroom but the "electronic classroom" remains in the realm of science fiction. As an educator who has watched students dose off during my lectures, I could use some form of electronic technology to shock my dozing students back from their world of dreams, but I really see little use for much sophisticated educational technology.

Decades ago when I helped pioneer the development of computer animation, some of us saw the potential for a computer-animated movie to help explain difficult concepts to students. But after years of waiting for the development of such technology, I now realize that I can do just as well with chalk, waving my arms, and some attempts at humor. The real challenge is to keep my students attentive and motivated.

The biggest success of tele-education has been the creation of popularized programs that educate millions as the shows are broadcast over the air on network television or over cable channels. I vividly still remember the "Bell System Science Hour" and "Omnibus" from the early days of television. "Victory at Sea" is still going strong over four decades after its premiere in 1952 on Sunday afternoons. Today's PBS educational programs demonstrate that education can be both entertaining and informative at the same time.

TELEMEDICINE

Telemedicine is the use of telecommunication in the practice of medicine. The use of video in telemedicine enables physicians to see patients remotely. With video telemedicine, specialists at a large medical facility can telediagnose patients and supervise nurse practitioners at a remote clinic. Interest in video telemedicine began in the early 1970s. The federal Department of Health, Education, and Welfare funded a number of trials of video telemedicine at that time. The AT&T picturephone system installed in Bethany/Garfield Hospital was one of the trials supported by HEW. Many of the telemedicine trials were mostly technology demonstrations with little or no scientific evaluation. However, some of the trials performed in the 1970s did evaluate the effectiveness of the video compared to audio-only in controlled experiments and are summarized below.

Three neighborhood health satellite centers were connected to Cambridge Hospital in Massachusetts by a one-way video link with two-way audio. The video was then compared with the audio-only telephone for about 350 patients for consultations between the nurse practitioners at the satellite centers and physicians at the hospital. The video consultations took longer than the telephone consultations, and the telephone was found to be as good as the video. However, the video consultations resulted in fewer immediate referrals to the hospital. The experimental design in this trial was quite thorough.

In a study supported by NASA, high-quality video-taped medical examinations performed by medical assistants were presented to physicians in a controlled experiment for their diagnoses. Otoscopes, ophthalmoscopes, and other instruments were used as necessary. The audio-only condition showed no significant difference in diagnostic success compared to the video systems. A number of technical video parameters were varied, but most had little net effect, and the commercial NTSC standard was adequate. This experiment confirms that a competent medical assistant can relay any visual information to a physician over a telephone. Similar results were found in a study performed at a City of Toronto community health clinic. The 600 patients in the study were telediagnosed over two-way television and two-way hands-free audio-only conditions. There were no statistical differences in diagnostic accuracy, consultation time, and referral rates to other medical personnel.

It seems that when no market can be found for some new technology, it is then proposed for use in either education or rural areas. In 1976, the National Science Foundation funded a report by the University of Michigan that not surprisingly proposed the use of telemedicine for rural areas. Yes indeed, video can be used to bring medical diagnoses to rural areas, but a nurse practitioner can use the telephone just as well to discuss the case with a distant specialist. And that same telephone line can be used by physicians to exchange medical records and data, for the transmission of patient monitoring of blood pressure and other vital signs, and for remote access to computerized medical records.

THE VIDEOPHONE

The videophone allows you to see the person you are speaking to on the phone. Of course, the other person sees you too in a two-way video telephone call.

AT&T interviewed 700 people who had used a demonstration picturephone at the 1964 New York World's Fair, and based on the results, went forward with the development and marketing of its videophone, called the picturephone—a project that would ultimately cost AT&T an estimated $500 million. In its 1970 Annual Report, AT&T predicted that there would be 50,000 picturephones in use in 25 cities by 1975. However, most business and residential customers could care less, and the AT&T picturephone became the Bell System's "Edsel."

The AT&T picturephone offered full motion, a monochrome picture, and about half the resolution of a conventional TV image. It worked over three phone lines, and though fairly costly, was somewhat affordable. The AT&T picturephone utilized sophisticated image technology and was a well designed product offering desk-top, impulse use.

After the picturephone's flop, AT&T entered a market research program to determine the market segments for two-way, switched, video communication. I transferred from Bell Labs to AT&T and became involved in this market exploration program, . My task was to peform independent evaluations and report to the manager in charge of the program. We explored the use of two-way, visual communication in health care, criminal justice, and teleconferencing.

The public-room intercity video teleconferencing service that we offered was at the height of the energy crisis, yet few people used it, even for free. An internal system for AT&T employees was used, however, though at somewhat lower usage than we had hoped. The picturephone system we had installed in Bethany/Garfield Hospital in Chicago was removed the day that federal support ceased. Our market research indicated that it was being used mostly as a hot-line telephone system rather than as a face-to-face visual telephone. The picturephone system we had installed in Phoenix for use in criminal justice so that public defenders could "meet" over the picturephone with their jailed clients was indeed used successfully, but the switched feature was not needed:

a point-to-point closed-circuit two-way video system would have worked just as well. In the end, we had to admit defeat in our efforts to find a market for switched, two-way, visual communication.

VIDEOTEX HOME INFORMATION

After spending well over $50 million for development and other expenses, the British Post Office launched a new service in the late 1970s called viewdata, or more generically, videotex. The British viewdata service enabled the home consumer to use a special terminal to access a wide variety of information from a central data bank over a telephone line. The home TV set was used to display the information which was organized in a tree fashion for access using a simple key pad. Color and simple graphics made the display pleasing to the eye.

In 1978, the British Post Office forecast one million users by the end of 1981. The actual number of home users was only one thousand. The failure of the British viewdata service did not prevent others from replicating it.

In 1980, AT&T and Knight-Ridder Newspapers, Inc. conducted a trial, called Bowsprit, of a videotex service. Knight-Ridder formed a new company, Viewdata Corporation of America, to design and develop the videotex service, called Viewtron. The data base contained about 15,000 frames of information. AT&T designed and provided the home terminals; Southern Bell provided the network. A few dozen terminals were circulated among 125 homes in Coral Gables, Florida over a six-month period starting in mid July 1980 and lasting to the end of January 1981. The terminal, which accessed the service over a phone line, had a full alphanumeric keyboard, just like a personal computer, but used the home TV set for display.

The participants in the trial were chosen to be positive toward the concept of accessing information over a home terminal. Based upon the positive response of the trial participants, AT&T and Knight-Ridder decided to launch the service commercially, staring with a market test in southern Florida in 1983. The consumer response was dismal, and in 1986, Knight-Ridder abandoned Viewtron, and today admits to having lost $50 million. I would estimate the loss to AT&T to be a few times that amount.

Although the participants in the trial had little interest in accessing the large data base that Knight Ridder had created, they did use the system to send text messages to each other. This finding is consistent with today's use of the Internet and other data networks to send text messages, called electronic mail or e-mail, from one person to another. The surprise to me about the trial was how quickly the participants discovered the potential for sending messages to each other since we had not promoted this application very much. Furthermore, the terminals were rotated every few weeks which gave the participants little time to discover for themselves this well-used application. Clearly it was very useful and important to them.

Although the British viewdata system used a system of simple block mosaics for graphics, AT&T and VCA were convinced that much fancier graphics and color were essential to make the system user-friendly and entertaining. Hence, Viewtron used a more complex and flexible graphics system, called NAPLPS (for North American Presentation Level Protocol Standard). This costly and complex system greatly added to the cost of the terminal and also to the complexity of the data base. Later, when teletext was considered for North America, NAPLPS was proposed, and its cost and complexity undoubtedly are a factor in why the United States still does not have any teletext service. But more about these and other missed opportunities in a later chapter.

AT&T published telephone books and thus had a keen interest in electronic directories and electronic classified advertising, known as electronic Yellow Pages. In 1979, AT&T conducted a trial in Albany, NY of such an electronic information service, called EIS-I. A total of 75 homes and 8 businesses participated in the trial. White pages were accessed 40 percent of the time by residential participants and 60 percent by businesses. Yellow Pages were accessed 19 percent of the time by residential participants and only six percent by businesses. The AT&T directory organization was excited by the prospects for its electronic information service, but impatience and greed later killed this opportunity, as we shall see in a later chapter.

A few years after the start of the Knight-Ridder project, AT&T became involved as the supplier of videotex terminals to the Times Mirror Company in its trial of a videotex system in Orange County, California, conducted in 1982. A commercial videotex service, called Gateway™, was launched in 1984 for Southern California by Times Mirror. The service, designed by Times Mirror in partnership with Infomart of Toronto, was nearly identical to that offered by AT&T with Knight-Ridder. Plans were developed to introduce the service into about four other Times-Mirror newspaper markets in the United States. In 1986, Times Mirror admitted that the consumer reaction to the videotex service "was not sufficient to warrant full-scale development of Gateway as an ongoing business" and the service was discontinued.

My personal experience with Gateway is most telling. One of the graduates of the university where I taught in Los Angeles invited me to visit the headquarters operation for the trial. I remember that I was very impressed with all the energetic, enthusiastic, well-dressed young people and also the impressive computer system. I started to think that my negativism and doubts toward videotex were all wrong. At the end of my visit, I asked whether there was a Bullocks department store in the mall across the street so that I could return an item I had purchased at the same chain at a different location. My host said that we could use the videotex service to find out. Her fingers did the walking across the keys of the terminal as we searched frame after frame of information. After nearly an hour of frustration, we finally used the phone book. My negative suspicions of the future of videotex had been confirmed. Interestingly, it was only a few months after my visit that Times Mirror discontinued the service.

In 1994, Bell Canada closed the doors of its videotex service, called Alex. Alex offered such services as home shopping, travel information, reservations, home banking, and e-mail. Bell Canada's Alex started in 1990 and initially had over 150 information providers. As has occurred again and again, real consumers cared little about an electronic data base of home information and used Alex mostly as a form of e-mail for chatting.

All these past experiences with videotex confirm the conclusion that a single electronic data base that promises all the world's information at your fingertips simply is unable to deliver on that promise. The data bases are not user friendly, and it is usually impossible to find anything. Most people obtain their information from other people—hence the success of e-mail.

The consumer response to videotex services that promise all sorts of information at your finger tips has been consistently negative. Nevertheless, confirming once again that market predictions are nearly worthless, in 1983 the business consultant firm Booz, Allen, and Hamilton, Inc. forecast 17 to 30 million households in the United States would have videotex by 1993. However, the decentralized approach of the Internet to electronic information does seem successful.

The electronic newspaper is so alluring that the past losses do not seem to bring any sense of reality. Knight-Ridder and Times Mirror are back into the electronic newspaper, now offering it on-line over the Internet and Prodigy, and promising a host of futuristic services with rhetoric almost identical to that of their past flops. While there is no doubt that students and scholars use electronic data banks in their research, I continue to doubt that other than such highly specialized uses there is a mass market for the electronic newspaper. But newspapers have deep pockets and will continue to whip each other into a frenzy in their search for the secret to the electronic, on-line newspaper.

INTEGRATED SERVICE DELIVERY

In 1979 while I was at AT&T, I suggested the development of an integrated local distribution system that would satisfy the needs for voice, data, and video communication. I foresaw the possible use by the late 1990s of optical fiber or other broadband transmission medium to carry an integration of voice, data, and video services. I also suggested the interim use of a single pair of copper wires to connect many homes for data communication, what I then called a shared data line and what today is called a local area network, or LAN.

These ideas of mine were consistent with technology trials of optical fiber for local telephone service and CATV service that were occurring in Canada in the late 1970s and early 1980s. In 1978, Bell Canada used optical fiber to provide telephone service to 34 homes in the Yorkville section of Toronto. Manitoba Telephone Systems was planning a trial of telephone service and video over coaxial cable in the same time frame. Other trials were planned in France (Rennes), in Japan, and in Germany. I noticed the lack of a Bell System presence in the investigation of these new broadband technologies for local delivery of telephone, data, and television signals. I do not know the final outcome of all the non-Bell trials, but my guess is that the technology and market were not yet then ready. But, a few years later, the concept of an integration of services again arose.

ISDN AND THE SUPERHIGHWAY

In the mid 1980s, the concept of integrating voice, data, and image signals all in a common stream of digital bits first appeared. The concept was called an Integrated Services Digital Network, or ISDN for short.

The basic idea was a network that was totally digital from one end to the other. A digital outlet in homes and offices would connect to the ISDN, which would require special digital telephones. High speed data communication at 16 kbps and 64 kbps was promised. The problem was that a costly reconstruction of the telephone network was required along with new telephones. The modems that are used today for data communication over the telephone network have so increased in speed that they apporach rates nearly as high as those promised with ISDN.

The need for an ISDN evaporated, although we may indeed some day have a network that is completely digital end-to-end. Along the way, broadband video was added to the list of services to create the concept of a Broadband Integrated Services Digital Network, or BISDN, which was shortened to Integrated Broadband Network, or IBN for short. A BISDN somehow combined with the Internet seems to be what the information and telecommunication highway means to some of its proponents.

TECHNOLOGY FOR SOCIAL NEEDS

In 1971, my career took a major turn, and I left Bell Labs to join the staff of the White House Science Advisor where I worked for the next two years on science and technology policy. The supersonic transport (SST) project had just been terminated and attempts were being made to focus the aerospace industry on national social problems.

In 1971, the White House initiated an investigation of new technological opportunities to solve social problems through the application of technology. An initiative coordinated by the National Aeronautics and Space Administration (NASA) focused on communication for social needs. The report of the investigation was issued in September 1971 and recommended the use of communication technology to solve problems in such areas as education, health services, law enforcement, postal service, and disaster warning. Not surprisingly, given the leadership of NASA in conducting the study, there was much emphasis on the use of communication satellites, although broadband coaxial cable was also recommended.

This project was my first exposure to technology push by the federal government on a grand scale. The communication initiative was only a small portion of a much larger effort. In the end, the good sense of the Office of Management and Budget prevailed, and the entire multi-billion dollar New Technological Opportunities Project (NTOP) was shelved. Indeed, "history repeats itself." Today, with the elimination of the cold war, the aerospace and defense industries are again attempting to discover ways to use technology to solve social problems. My guess is that the final outcome will be the same as yesteryear's NTOP.

LESSONS FROM THE PAST: SPIRAL OF HYPE

The experience in two-way CATV in the early 1970s in the United States excited the Japanese to perform their own trials of wired cities, even as the experience in the United States was failing. The Japanese activity then rekindled interest in the United States. All this became a spiral of hype in which each country excited the other in a frenzy of copy-cat activity.

There are many lessons to be learned from this history of failed attempts to use interactive and two-way video to solve various social problems. Most social problems have many components, and it should not be a surprise that technology is not a sweeping solution. For example, medical care must be carried out under the supervision of a physician. This can be interpreted to mean that a physician must "see" subordinates to supervise them and that therefore a video link would facilitate this supervision over distance. However, a telephone call could be just as effective in such supervision and much less costly and complicated.

Much of education involves students interacting with other students, and this interaction occurs most easily in the classroom. Students need to be motivated to learn. Indeed, technology in the form of a video classroom or computer-assisted instruction can motivate students to learn, but so too can an effective instructor in a conventional classroom setting which also allows students to interact directly with each other.

One problem with many of the evaluations of interactive media described in this chapter is the so-called "Hawthorne effect." Decades ago, a study was conducted to improve productivity at the Western Electric plant in Hawthorne, Illinois. As various physical aspects of the work environment were improved, productivity increased as you would expect. But then the physical aspects were changed back to the less appealing condition, and productivity continued to improve! The simple fact that someone seemed to care and was attempting to improve the work environment was sufficient to improve productivity rather then the actual environment. What this means is that simply performing a study involving some new technology is sufficient to improve performance, regardless of whether the technology is really an improvement or not.

In the next chapter, we continue our study of the past, but with a focus on more recent past events and recent activities.

READINGS

Bashshur, Rashid, "Rural Health and Telemedicine," NSF Grant No. GI-41770, School of Public Health, University of Michigan, Final Report 1976.

Clark, Richard E., "Reconsidering Research on Learning from Media," *Review of Educational Research*, Vol. 53, No. 4 (Winter 1983), pp. 445-459.

Davis, Jerry G., "Video Requirements for Remote Medical Diagnosis," SCI Systems, Inc.: Houston, TX, Final Report, June 1974.

Gottschalk, Jr., Earl C., "Firms Are Cool To Meetings By Television," *The Wall Street Journal*, July 26, 1983, p. 35.

Kay, Peg, "Social Services and Cable TV," The Cable Television Information Center, July 1976.

Lucas, William A., Karen A. Heald, & Judith S. Bazemore, "The Spartanburg Interactive Cable Experiments in Home Education," Rand: Santa Monica, CA, Report R-2271-NSF, February 1979.

Moore, Gordon T., Thomas Willemain, & Rosemary Bonanno, "Comparison of Television and Telephone for Remote Medical Consultation," *New England Journal of Medicine*, Vol. 292, No. 4, April 3, 1975, pp. 729-732.

Moss, Mitchell L. (Editor), "Two-Way Cable Television: An Evaluation of Community Uses in Reading, Pennsylvania," New York University, April 1978.

Noll, A. Michael, "Teletext and videotex in North America: Service and system implications," *Telecommunications Policy*, March 1980, pp. 17-24.

Noll, A. Michael, "Videotex: Anatomy of a Failure," *Information & Management*, 1985, pp. 99-109.

O'Neill, John J., Joseph T. Nocerino, & Philip Walcoff, "Benefits and Problems of Seven Exploratory Telemedicine Projects," The Mitre Corporation, February 1975.

Park, Ben, "An Introduction to Telemedicine," The Alternate Media Center, New York University, June 1974.

Reich, Joel J., "Telemedicine: The Assessment of an Evolving Health Care Technology," Washington University: St. Louis, MO, Report Numbers R(T)-74/4 & THA 74/6, August 1974.

Rockoff, Maxine, "An Overview of Some Technological/Health-Care System Implications of Seven Exploratory Broad-Band Communication Experiments," *IEEE Trans. Comm.*, Vol. COM-23, No. 1, January 1975, pp. 20-30.

"Communications Technology for Urban Development," Committee on Telecommunications, National Academy of Engineering, June 1971.

"The Feasibility And Value Of Broadband Communications In Rural Areas," United States Congress, Office of Technology Assessment, April 1976.

Chapter 4

HISTORY REPEATS ITSELF: LESSONS FROM TODAY

One would think that a twenty year history of failed attempts to find gold at the end of the broadband telecommunication rainbow would be sobering enough to caution others from such fruitless searches. However, the communication industry seems to refuse to learn from its past mistakes. And today we find many CATV firms and telephone companies again in hot pursuit of the superhighway Holy Grail. A few skeptics have seen through the futility of this search and have sarcastically called the information superhighway the "highway of hype" or the "superhypeway." In this chapter, we will review some of these more recent events and the lessons we should learn from them.

GTE CERRITOS PROJECT

About 10 years ago, GTE (a local telephone company) and Apollo CableVision (a CATV company) joined forces to provide CATV service to the residents of Cerritos, California. This was truly an exciting marriage, since traditionally telephone companies and CATV companies are enemies of each other. Now, ten years later in 1996, Apollo abandoned its efforts and sold itself to GTE. How did this partnership form and what tore it asunder?

Cerritos is a small community of about 15,500 households just south of Los Angeles. It did not have CATV mostly because the city required the cable to be buried underground which was too costly to attract most

CATV firms. All that changed in 1985 when GTE and Apollo CableVision became involved in a joint venture to wire Cerritos with an advanced cable TV system.

As a city, Cerritos promotes itself as a "step ahead with some vision" and hence was interested in promoting the use of advanced technology for its CATV system. GTE wanted to investigate the use of broadband fiber to carry a variety of "new" services to the home, such as video on demand, home information, and two-way video telephones. From GTE's perspective, the Cerritos facility was intended for technology tests and service experiments. From Apollo's perspective, the Cerritos system was intended to provide conventional CATV service.

An affiliate of Apollo had developed the technology for very quickly and inexpensively trenching a city street to install a conduit carrying optical fiber along with conventional coaxial cable to the homes in Cerritos. The trenching and installation was done so smoothly and efficiently that few people even knew that their street had been trenched, cable laid, and repaved all in a day. The cable had the capacity to carry 78 channels. GTE paid the $12 million that the installation cost, and initially leased half the capacity to Apollo. The Apollo half was used for conventional CATV, and the remaining half was retained by GTE for its use to test advanced services. The installation of the system was completed in 1988.

The first problem that arose between GTE and Apollo involved the set-top convertor boxes, inside wiring, outside drop wire, and billing system, all of which were owned by Apollo. GTE wanted to own these items and include them in the leasing arrangement with Apollo. Ultimately, most of this was done, but Apollo retained the billing system and in return obtained agreements from GTE for right of first refusal for the remaining capacity on the system and also for 22 years of no competition.

The next problem that arose a year later was over the terms of the lease. The rates set by GTE seemed too high to Apollo, and GTE refused to negotiate. Apollo then exercised a prepayment option and obtained its own financing to prepay the 15-year lease for its half of the capacity.

Policy issues arose since GTE was in partnership with Apollo and, in effect, was providing CATV service. Federal cross-ownership rules in effect at that time forbade a telephone company from providing CATV service. Although in 1988 the FCC waived these rules for five years for the Cerritos trial, the courts later ruled that they applied in this case. But

Bell Atlantic had won a battle with the courts to allow it to offer CATV service, and this excited GTE about similar prospects for the superhighway in its territory. In the meantime, the Cerritos trial had come to an end, and it looked as if GTE was going to leave Cerritos. Apollo wanted to lease GTE's half of the capacity to provide additional programming to Apollo's CATV subscribers. GTE offered its half to Apollo for $95,000 per month, which seemed far too high to Apollo. Perhaps GTE wanted to price Apollo out of Cerritos since GTE had now come to believe in the superhighway and wanted the entire system for its own purposes.

GTE was left somewhat empty-handed with its 39 channels of unused capacity of the system and was seeking ways to use it. For a while, Apollo provided conventional CATV service over its 38 channels to the residents of Cerritos. GTE had its 38 channels and provided a videotex service, called Main Street™, to a few hundred homes and pay-per-view TV service, called Center Screen™, to a few thousand homes. Rumor was that neither of the GTE services was very successful. However, Apollo believed that GTE's pay-per-view TV service was a violation of its non-competition agreement with GTE and thus sued GTE. In the end, Apollo settled by selling its business to GTE, and GTE now owns the entire CATV system for Cerritos.

GTE promoted Cerritos as a trial and technological test-bed for all sorts of new two-way broadband services, ranging from video on demand to two-way video telephony. A fancy customer demonstration center was created by GTE in Cerritos to demonstrate many of the new services, although some people considered GTE's attempts little more than "smoke and mirrors" and a showcase to politicians and regulators For example, GTE's approach to testing video on demand involved only two families and a bank of VCR machines at the GTE demonstration center. The various services that GTE has tested or is testing currently also include educational video classrooms, telephone service over optical fiber, and videotex-like home information. One problem was GTE's continuing refusal to disclose its real intentions in Cerritos or the results of its trials there.

Indeed, as many people have observed, the GTE project in Cerritos bombed in terms of new interactive services. The real success in Cerritos was Apollo CableVision's trenching operation and the use of optical fiber in the CATV backbone network. When GTE set rates to sell capacity on the broadband network to Apollo, the rates were too high to be competitive, just like when AT&T provided CATV capacity in the 1960s.

The emphasis on technology by the city of Cerritos is similar to the techno-image that Reston, Virginia sought decades ago with TICCIT. Cities know little of telecommunication technology, and these kinds of attempts to gain techno-publicity ultimately backfire. Cities would be wise to stick to providing conventional services and to take a low profile when experimenting with new technology.

The lesson to be learned from the GTE project in Cerritos is that like cats and dogs, cable companies and telephone companies do not get along well together. Fights and battles are the result. But it is difficult to win in a battle with a giant company like GTE, and so in the end, GTE got its way and now owns the Cerritos CATV system. Perhaps that is what GTE wanted from the beginning.

THE VIDEO CLASSROOM

One application that GTE pursued in Cerritos was the provision of educational programming to two local elementary schools in Cerritos by linking the schools with two-way video. Indeed, video can be used to link students in remote classrooms, and for decades many engineering schools around the country have been broadcasting their classes to remote sites. There was nothing new here at all, although GTE seemed to think so.

Linking two local schools by video is little more than old-fashioned closed-circuit TV (CCTV) and does not utilize the switching that makes telecommunication so unique compared to CATV. Bell Atlantic has recently connected a number of schools in New Jersey with two-way video at 45 Mbps. The system enables schools to share courses, and undoubtedly is effective. As we saw in the past chapter, students can indeed be taught over video without having to travel to a class to be present physically.

One real issue in the use of video in education continues to be cost. Bandwidth is costly, particularly when provided by the telephone company. I doubt that local schools will ever pay what the bandwidth and system should actually cost; the schools will want it for "free" or at subsidized rates. The Telecommunications Act of 1996 allows the subsidization of such low rates for schools.

As we have learned over and over again, the use of technology in education is very much "apple pie and motherhood"—until real money becomes involved. The problem for the phone companies is that if video is provided at rates below cost, the schools will be motivated to resell capacity to potential competitors of the phone companies.

BELL ATLANTIC AND SAMMONS COMMUNICATIONS

In 1992, the state of New Jersey passed a Telecommunications Act which replaced rate-of-return regulation with price-cap regulation of Bell Atlantic in return for Bell Atlantic's promise to accelerate the deployment of advanced technology in New Jersey's telecommunication infrastructure. This new infrastructure would utilize optical fiber extensively in the local loop, bringing fiber to the curb outside each home. Bell Atlantic believes in integration of telephone service with video so that a consumer can dial up a centrally located data bank of videos, a service known as video dial tone.

In the Fall of 1992, Bell Atlantic joined forces with Sammons Communications to offer video dial tone service to about 11,000 homes in Florham Park, New Jersey in a technology and market trial. Bell Atlantic would own and maintain the fiber system and lease capacity on it to Sammons. This common-carrier CATV model is similar in philosophy to the CATV systems owned by AT&T in the 1960s and also to the original planned relationship between GTE and Apollo CableVision in Cerritos. The Bell Atlantic system would utilize fiber to the curb to provide a variety of integrated switched broadband services. Since the system was not installed, what happened?

One problem was that the technology was not deliverable. Far more had been promised and designed on paper that was ever available "off the shelf" for real installation and use. Another problem was that Bell Atlantic was not able to obtain regulatory approval from the FCC to offer the service. And lastly, the rates proposed for evaluation to Sammons to lease capacity appeared to exceed standard operating costs, although Sammons would have saved in maintenance expenses and in capitalization costs. In the end, Sammons could wait no longer for Bell Atlantic, since Sammons had scheduled a rebuild of its existing CATV system in the city. Bell Atlantic and Sammons severed their relationship amicably in July, 1994.

Bell Atlantic's promotion and hope for video dial tone has been intense. At one time, Bell Atlantic predicted it would have 1.2 million homes connected by the end of 1995. In January 1993, Bell Atlantic was reported in an article by Charles F. Mason to have predicted yearly revenues in 1995 of $1,754 per subscriber from video dial tone which would rise to $9,827 per subscriber by the year 2000. However, the average monthly CATV bill in 1994 was about $33 per subscriber, or about $400 per year. Bell Atlantic proposes to increase this bill four-fold almost immediately! It is unbelievable that consumers would place such a high value on video on demand, or would even have the discretionary income necessary to afford such a luxury. Clearly, this is yet another example of wishful thinking coupled with superhighway superhype and nonsense.

The Bell Atlantic initiative in New Jersey is known as "Opportunity New Jersey," and if Bell Atlantic's revenue projections for video dial tone are correct, then perhaps the initiative should more appropriately be called "Opportunity Bell Atlantic."

Although the Sammons deal has fallen through, Bell Atlantic is pursuing a video dial tone service offering in Dover Township, New Jersey. Roughly 40,000 homes will be connected to the system, using technology developed by Broadband Technologies, Inc. The technology was tested in late 1995 for a few hundred homes, and a commercial launch of the service was planned for early 1996 with nearly 15,000 homes to have initial access. Bell Atlantic will act as a common carrier for the video service and will sell space on the system to video providers. Nearly 400 video channels will be available. How well the service works out for Bell Atlantic and the video providers remains to be seen.

BELL ATLANTIC AND DOVER TOWNSHIP

In May 1996, Bell Atlantic gave me a tour of their video dialtone facility in Tom's River, a community in Dover Township, New Jersey. A total of about 4,500 homes were connected to the fiber system, with 6,000 to be connected by June.

The video programming is being provided by FutureVision, a company formed about three years ago to provide video programming in the Dover Township service launch. I visited FutureVision's head-end as part of my tour. FutureVision provides programming on about 96 of the total of 384 available channels. FutureVision digitizes and compresses its video programs using MPEG-2 at six Mbps and then transmits the multiplexed digital signals to Bell Atlantic's central office. FutureVision has about 2,000 subscribers and is engaged in a fierce price war with the existing, conventional cable company. There were many costly digital convertors to perform the digitization and compression. FutureVision told me that their head-end facility cost about $10 million.

The system was demonstrated to me at the FutureVision head end. I was surprised by the one-second delay that it took to change a channel. But to change a channel, the set-top box (manufactured by Phillips) must send a signal all the way back to the central office where a video switch selects the requested channel. All this apparently takes time.

I was shown an interactive show on the Food Network. The interactivity was simply to request to see a food menu on the screen or to chose to have the menu sent by fax. The on-screen food menu was difficult to read since a cooking show was in the background. I asked whether the show could be eliminated, but it could not. Apparently, the centralized video switching makes this difficult to do.

I saw Optical Network Units (ONUs) mounted on telephone poles and was told that some ONUs are buried underground in some neighborhoods. The manager in charge of the project had a team of installers in telephone trucks wiring neighborhoods and installing fiber, ONUs, and set-top boxes. I was told that no technical problems had occurred thus far.

In a later chapter, we will examine the technological uncertainties and issues associated with video on demand and video dial tone. In particular, the Dover Township system, although the technology works, is an awkward and costly approach to providing video.

VIDEO ON DEMAND

In 1995, Rochester Telephone Corporation, now called Frontier Corporation, halted its market test of video on demand in Brighton, New York. The system provided access to about 100 digital videos to 52 homes and was priced from 50 cents to $4 per video. The service was not viable commercially. It was reported that consumers in the trial still went to the video store.

Southern New England Telephone (SNET) claims subscribers will pay for video on demand, but nevertheless has pulled back on its plans citing technical problems involving digital-video file servers.

If video on demand were such a great idea, then audio-on-demand which is technically more feasible would already be here today. It is not because consumers have little need for such a service. Much of the fun in listening to the radio is not knowing what music will be played next. And much of the fun in buying CDs is browsing through the bins.

At the writing of the present book, the superhighway seems to be crumbling. Many trials around the country are being delayed, postponed, or canceled. The media frenzy about the superhighway also seems to be waning with less and less press coverage each day. But some hold-outs and a few late entrants continue the promotion and hysteria. Not surprisingly, education has now discovered the superhighway, perhaps smelling publicity, government hand-outs, and new sources of funding.

THE ELECTRONIC CLASSROOM

The grammar-school classrooms of the nation—and ultimately their closets of unused technology, it seems—have been stuffed with film strip projectors, audio tape players, movie projectors, television receivers, and today computers, but yet we feel that something is still wrong with education. Technology has not solved the problems of education, yet the search for the electronic classroom continues.

Education is much like apple-pie and motherhood, and hence it is difficult to oppose, or even question, the use of technology in education. But the history of educational technology is strewn with failure to improve and change radically the quality of education! I remember the promise of computer animation to assist the teacher in presenting complex material; of video delivery of courses to distant locations around the world; of databanks storing all the world's knowledge to be accessed and used by students. Given all this technology, and more, the teacher in front of the classroom is still the major way most students are taught. So, is the electronic classroom just another attempt in a long series of attempts to solve the problems of education through technology?

Clearly something is wrong with our educational system. I routinely have graduate students who have difficulty in completing an English sentence and who can not do simple arithmetic. Indeed, something went wrong back in elementary school. If these students slept their way through grammar school—which I doubt—then some form of electronic classroom, such as I mentioned earlier in jest, that shocks awake sleeping students might be an appropriate use of technology in the classroom!

Does technology have a place in education? Yes, indeed, today's technology has already had major impact in education, particularly at the higher levels. Computers have streamlined much of the administrative bureaucracy. Students at many universities now register from home and from dormitory rooms using a touchtone telephone. The use of electronic databases for research is routine. The copy machine has saved journals from theft and destruction. Computers used as word-processing typewriters help students produce more legible term papers. Copy machines enable teachers to bring the most recent material to class to distribute to students as readings. The overhead projector shows visual graphs to the class. Students bring audio cassette recorders to class to tape lectures.

Students will continue to use personal computers; computers and communications will continue to improve the delivery of administrative services; schools will continue to use video to deliver courses to distant locations; and local area networks will continue to link campus computers to share resources. But this day-to-day technology seems far from the wonders of tomorrow's electronic classroom, whatever it may be.

Education is a complex and huge system involving such factions as teachers, students, administrators, family, funding agencies, politicians, and public policy. The use of technology in the classroom is a very small portion of this system, although even a small incremental improvement can be worthwhile.

I believe that one key ingredient in quality education is motivation—a motivated teacher and students motivated to learn. If the teacher does not know the topic or does not care about the students, if the students are interested only in grades or buying a degree and not in learning the material, if the students are not held responsible for the acquisition of actual knowledge and skills, then quality education will not be the result. Indeed, there might be potential for educational technology to save money, but most schools would probably be better off spending the money in acquiring additional teachers or on improving the skills of existing teachers.

THE ELUSIVE ELECTRONIC NEWSPAPER

For the past fifty years or so, the newspaper industry has been searching for new ways to deliver newspapers to our homes and businesses using electronic transmission and replacements for paper. This search has thus far been fruitless and probably will continue to be so for the next fifty years!

Fifty years ago, newspapers were sent by broadcast radio to tens of thousands of home facsimile machines costing from $50 to $100. The *Chicago Tribune, New York Times, Miami Herald,* and *Philadelphia Inquirer* all transmitted facsimile editions. By 1950, the fax newspaper had disappeared. The introduction of television at the end of World War II was certainly a factor in defeating these early attempts at electronic newspapers.

In the early 1970s, the use of microfiche along with a small, hand-held, flat-screen viewer was explored for the delivery of a paperless newspaper. It never got off the ground. In the early 1980s, electronic "videotex" newspapers that would be delivered over telephone lines and displayed on home TV sets were developed by Knight-Ridder and Times-Mirror. They too bombed.

Today, interest has again returned to the electronic newspaper. Recycling will be eliminated and forests saved. Flat high-resolution screens, microprocessors, and CD-ROM are some of today's technologies that are relevant to the renewed enthusiasm for the electronic newspaper. Access will be near instantaneous, and the information accessed will be customized to the interests and information needs of each individual user.

Yet, the reality is that the technology is costly; the displays are not very readable; users are not able to articulate their information needs; and the batteries need to be recharged. For packing vast amounts of high-resolution information into a readable and easily accessible medium, paper continues to be nearly impossible to beat.

One would think that there were some lessons for the newspaper industry from this history of failure of the electronic newspaper. Yet, newspaper publishers continue to fear that someone—perhaps the Baby Bells—will stumble upon the secret of the thus-far elusive electronic newspaper.

As a response to these thus-far unfounded fears, the newspaper industry continues to lobby to keep the Baby Bells out of the information industry. The Telecommunications Act of 1996 prohibits the Baby Bells from electronic publishing for four years from the enactment of the Act in February 1996. Perhaps there is reason for some fear since computer-based technology could be a great replacement for classified advertisements. NYNEX, for example, has joined forces with Prodigy to offer a computerized form of electronic Yellow Pages which could offer timely information and thus compete with the classified ads in newspapers.

Such "on-line" services as Prodigy already offer news in addition to advertisements and access to all sorts of computerized information over the Internet. Indeed, the electronic newspaper is already here. And if it is as wonderful as its proponents tell us, why aren't the newspaper publishers shaking in fear? Can it be that they know that we all in our hearts continue to have a fondness for the paper and print of the newspaper? It is already portable, recyclable, easily readable, instantly accessible, and contains all the information one would ever wish.

An electronic news service that I could use would be a weekly listing of classical music concerts. Now, each Thursday I rush to pick up the free weeklies for their listings of music events. An electronic version that I could access over the Internet would save me much trouble and would be more inclusive and extensive. I would still read the weeklies, though, for their interesting articles.

I recall the paperless office that was just a few years away and that seems to continue to be forever just a few years away. All of our personal computers and laser printers, rather than taking us closer to the paperless office, simply enable us to produce reams of drafts, all on paper. It seems that paper will be the medium well into the twenty-first century, and the quest for the electronic newspaper must join the paperless office and picturephone as yet another of today's technological Holy Grails.

TELEMEDICINE RETURNS

Telemedicine is back again, this time over the superhighway! Remote two-way video consultations and video teleconferencing of medical experts are some of the promised "new" uses of the superhighway. The transmission of medical records and data are also proposed. Indeed, the recording of medical records and images in digital form on appropriate digital media rather than on paper and film makes much sense. However, standards are needed. It might be easier for a patient to carry a computer disk rather than paper and film.

What seems to needs more promotion is the opportunity to use existing technology more extensively to perform such activities as remote patient monitoring over telephone lines and the transmission of medical data by facsimile. I am told that pacemakers are today already being monitored over phone lines.

LESSONS FROM TODAY

Perhaps what is most upsetting about today's hysteria and hype about the superhighway is that most of it is virtually a verbatim repeat of the rhetoric of two decades ago. But the promotion of these over-promoted technologies has its own life cycle. We need think back only a few years to remember the "revolution" and great promises of such dreams as virtual reality, multimedia, and high definition TV (HDTV). Yet, these terms hardly appear anymore today in the popular press, and we wonder what has happened to these revolutions.

REVOLUTION OR EVOLUTION?

Very little progress in technology is truly revolutionary. Instead, much progress consists of incremental improvements—an evolutionary process. Yet, the term revolution is used again and again, usually unjustified.

The first flight in a winged aircraft by the Wright brothers was truly revolutionary, but after this innovation nearly all progress in aircraft has been evolutionary, such as: breaking the sound barrier, jet engines, and computer control. The initial invention of television was revolutionary, but such progress as color, improved cathode-ray-tube displays, and UHF was evolutionary. The first recording of sound by Edison on his cylinders was revolutionary, but magnetic recording, the long-playing record, stereo, and even the digital compact disc were all evolutionary advances in technology. The invention of the telephone as a means to convey human speech over great physical distances was truly revolutionary, but nearly all technological advances in telephony since then have been evolutionary. Broadcast radio was revolutionary, but FM and stereo were all evolutionary advances in the technology.

Evolutionary advances in technology, though not dramatic in themselves, can have dramatic impacts in such areas as improved quality and lower prices. The earliest transoceanic telephone call was revolutionary, but such further advances as coaxial cable and optical fiber have been evolutionary. However, the effect of these evolutionary advances in the technology of transoceanic telephone service have been dramatic improvements in quality and much lower costs thereby making transoceanic telephony much more common.

The term "revolution" has simply become so much hype and should be used sparingly. The people promoting some specific technology as revolutionary are the last people to make such a judgement. The final decision as to the revolutionary aspect of an innovation might need to wait years before the final impact can truly be judged. Edison's phonograph was not commercially exploited until years after its invention, and Edison thought the market for it was solely as a dictation machine.

Change can occur too rapidly thereby leading to big mistakes. Any innovation needs time to be adopted, to be improved, and to blend into society. The process from innovation to implementation can actually benefit from delay, since delay gives the time to think clearly and to evaluate thoroughly.

EXPERTS

I recall a study sponsored by AT&T and performed by the Institute for the Future in 1970 in which 210 experts were asked their views of the future. The polling of experts was conducted in an iterative fashion—a methodology known as a Delphi study. One prediction was that there would be over 2 million picturephones in 1985. There is a lesson to be learned from such wishful—and usually wrong—thinking by so-called experts.

Experts are great at describing today's technology. Such descriptions draw upon their technological expertise. Experts are poor at predicting tomorrow's technology and are even poorer at predicting tomorrow's uses and markets for technology. I remember the prediction made in 1977 by Ken Olson, president of Digital Equipment Corporation, that "There is no reason anyone would want a computer in their home." And Thomas Edison was wrong in not recognizing the superiority of alternating current (AC) over direct current (DC) for the distribution of electricity. Experts become advocates when asked to predict, and this advocacy frequently becomes Utopian. Experts thus predict what they wish to occur, and their advocacy blinds them to the factors other than technology that shape the future.

CHANGING DEFINITIONS: WHAT IS IT?

Videotex initially had a very clear definition of being a home information service that used the telephone line for access, the home TV set for display, text and graphics in color, and information organized in a tree structure. Over time, the definition of videotex changed and broadened to include text-only information, full-motion graphics and video, and even conventional telephone service. By expanding the definition this way, the advocates of videotex never had to admit failure. If the past is prologue to the future, then we should soon see the superhighway ultimately defined to be the already existing telephone network and thus claim success.

DIGITAL OVERLOAD

The term "digital revolution" is most certainly overused, particularly when most people probably have no idea what digital means.

The concept of digital is decades old. In 1948, Bell Labs' scientists Bernard M. Oliver, John R. Pierce, and Claude E. Shannon invented the use of pulse code modulation—what is today called digital—to carry telephone signals. The first implementation of their invention was in 1962 for the digital multiplex system known as T1.

Computers are digital machines; people are not. People are analog creatures who read books and newspapers on paper, listen to music from loudspeakers and headphones, and watch videos on TV sets. All these media must create analog information for ultimate use by people. In our rush down the digital superhighway, we must not forget the human user.

Markets are created by normal analog human beings, like you and me. Mass markets are rarely created by techies who embrace technology for its own sake. One example is the computer techies who seem to prefer to communicate with computers rather than directly with other human beings. But computer techies have created much of the interest about the superhighway. Guess now who is poised at the on-ramp?

ELECTRIC UTILITIES AT THE SUPERHIGHWAY ON-RAMP

It seems that everyone wants to become involved with the information superhighway, and proposals are being made that electric utilities should build it. Even Vice President Gore has accelerated onto the information superhighway and has suggested that the electric utilities should help build it.

Indeed, electric utilities currently wire our neighborhoods and homes. But these wires carry electric power, and electric utilities know little of the two-way, switched nature of telecommunication signals. Electric power involves megawatts at kilovolts; telecommunication signals involve milliwatts at millivolts. Electric power is transmitted at a single frequency of 60 Hertz; telecommunication signals are sent at a wide range of frequencies from hundreds to billions of Hertz. Just because electric utilities string wire on poles does not mean they should be involved in telecommunication. I, for one, would not want the company that brought us the Three-Mile-Island incident to provide telephone or CATV service. The single frequency of 60 Hertz is about all they seem able to handle.

Natural gas, water, and sewage are all carried over pipe. This does not mean that all three should be provided by the same company over a single pipe. But as we have seen, there is much nonsense along the superhighway. Much later in this book we will examine the success of today's existing telecommunication infrastructure as a rejoinder to the hype of the superhighway. But before doing so, we will examine in detail the five factors needed for a successful venture. We will start with technology.

TECHNOLOGICAL UNCERTAINTIES

Technology is a key factor in suggesting and creating new options for the future. The recent success of the compact disc demonstrates the considerable impact technology can have on an entire industry in a very short time. The world of entertainment would be archaic without electronic television and its continued advances in technology. These advances have made TV sets affordable to all. I can remember the first TV set that my parents purchased by monthly payments since it was so costly. Today's TV set has a much better image, is far easier to tune, is considerably more reliable, and costs far less.

It is tempting indeed to conclude that all sorts of new technological wonders will create the technological Utopia previously presented. However, although it is tempting to conclude anything is possible technologically, there are some technological uncertainties that weaken the technology link needed to create market reality and a viable business venture. The next chapter explores these uncertainties.

READINGS

Cauley, Leslie, "Bell Atlantic Asks the FCC to Suspend Two Applications for Video Networks," *The Wall Street Journal*, April 26, 1995, p. B6.

Cauley, Leslie, "Phone Giants Discover The Interactive Path Is Full of Obstacles," *The Wall Street Journal*, July 24, 1995, pp. A1 & A9.

DeGeorge, Gail and Veronica N. Byrd, "Knight-Ridder: Once Burned, and the Memory Lingers," *Business Week*, April 11, 1994, pp. 74-75.

Eng, Paul M. and Ira Sager, "Prodigy is in That Awkward Stage," *Business Week*, February 13, 1995, pp. 90-91.

Gibbons, Kent, "SNET Drops VDT Plan, Goes Cable Across Conn.," *Multichannel News*, Vol. 17, No. 5 (January 29, 1996), pp. 1 & 48.

Kawasaki, Guy, "Potholes along the Information Superhighway," *MacWorld*, July 1994, p. 274.

Landler, Mark, "Dwindling Expectations," *The New York Times*, December 18, 1995, pp. D1 & D10.

Lehmann, Yves, "Videotex: A Japanese Lesson," *Telecommunications*, July 1994, pp. 53-54.

Lippman, John, "Tuning Out the TV of Tomorrow," *Los Angeles Times*, August 31, 1993, pp. 1, A13-A14.

Mason, Charles F., "Bell Atlantic sees big bucks in video," *Telephony*, January 18, 1993, p. 13.

Noll, A. Michael, "The Collapse of the 'Information Superhighway': A Perspective on the Failed Telco/CATV Deals," *Telecommunications*, July 1994, p. 19.

Noll, A. Michael, "Rethinking the Digital Mystique," *Telecommunications*, Vol. 30, No. 2, February 1996, p. 43.

Rainey, James, "Wiring a High-Tech Small Town," *The Los Angeles Times*, March 5, 1996, p. B2.

Robichaux, Mark, "Highway of Hype," *The Wall Street Journal*, November 29, 1993, pp. 1, A7.

Stoll, Clifford, *Silicon Snake Oil*, Anchor Books: New York, 1995.

Van, Jon, "Bells likely to buy, not build, into cable," *Chicago Tribune*, August 26, 1993, Business Section, pp. 1, 3.

Van, Jon, "A huge bet on a new information age," *Chicago Tribune*, October 14, 1993, Section 3, pp. 1-2.

Ziegler, Bart, Robert D. Hof, and Lois Therrien, "Dial R for Risk," *Business Week*, November 1, 1993, pp.1 38-139.

"America's Information Highway: A hitch-hiker's guide," *The Economist*, December 25, 1993, pp. 35-36, 38.

"Electronic cities: Move over, Mickey," *The Economist*, March 9, 1996, pp. 28-29.

"Electronic pathways to art appreciation: The virtual museum," *The Economist*, March 23, 1996, pp. 88-90.

"How the people of Rochester saw the future and yawned," *The Economist*, February 25, 1995, pp. 63-64.

"Multimedia: The Tangled webs they weave," *The Economist*, October 16, 1993, pp. 21-22, 24.

"On-line newspapers: Hold the front screen," *The Economist*, February 11, 1995, pp. 54-55.

"U S West and Cablevision: This time it really might work," *The Economist*, March 2, 1996, p. 61.

Chapter 5

TECHNOLOGICAL UNCERTAINTIES

Science fiction has created the widespread belief that almost anything is possible technologically. I guess if we are willing to wait long enough most of these expectations will be fulfilled. However, from a realistic perspective, the technology must be available "off the shelf" if some new product or service is to be provided today. We shall see in this chapter that there are many technological uncertainties that must be resolved before the construction of the superhighway can begin. Many promoters have ideas on how to solve these uncertainties, but commercial "off the shelf" solutions are not yet at hand. One must remember that fancy, color brochures and plans on paper do not alone create reality.

A HERTZ'S A HERTZ AND A BIT'S A BIT

We saw in an earlier chapter that much of the superhighway is a resurrection of old ideas. In the early 1970s, it was first proposed to combine telephone service with television over a single broadband medium—coaxial cable. This idea seemed sensible because all analog signals occupy bandwidth, and once assigned an appropriate width of spectrum, many different kinds of signal could all be carried over the same coaxial cable, under the philosophy that "a Hertz's a Hertz." The combining of signals was to be accomplished through appropriate frequency-division mutliplexing. The specific signals needed at each home would be obtained through appropriate demultiplexing. Any return signals from the home would be multiplexed into appropriate channels on the cable for transmission back to the head end.

Today's idea is to combine telephone service with television over a single broadband medium—most likely optical fiber. This idea has developed because all signals can be converted to digital format and combined through time-division multiplexing, under the philosophy that all bits are the same, or "a bit's a bit." This idea of combining telephony service with television over a single medium failed in the 1970s and will most likely fail again because there is much more to the nature of telephone service and television than only Hertz and bits.

A PICTURE'S WORTH A THOUSAND WORDS

One important difference between the video signals of television and the voice signals of telephony is bandwidth. As shown in the table, video requires about 1000 times the bandwidth of voice, regardless of whether the signals are analog, digital, or compressed. The old adage that a picture is worth a thousand words is confirmed.

	BANDWIDTH		
	Video	Voice	Ratio
Analog	4.5 MHz	4 kHz	1000:1
Digital	60-80 Mbps	64 kbps	1000:1
Compressed Digital	1.5 Mbps	1.2 kbps	1000:1

But capacity is only one dimension of the differences between television and telephony, and in particular, between CATV service and telephone service.

CATV VERSUS TELEPHONY

Television is a one-way medium. No matter how loudly you shout at the TV set, your words will not be heard by the broadcaster. Telephone service is two-way. You both listen and speak. Telephone service is interactive in the sense that what you say affects another person and then what they say to you in response. Television is not interactive—unless clicking the remote to change a channel is broadly considered interactive. These and other differences are shown in the table below.

	Cable Television	Telephony
modality	video	voice & data
directionality	one-way	two-way
bandwidth ratio	1000:1	
holding time	long (hrs)	short (mins)
simultaneity	many users	few users
purpose	entertainment	information
network	broadcast	switched
control	conduit & content	conduit only
medium: **- today** **- future**	coaxial cable (copper wire) fiber + coax	twisted pair (copper wire) fiber + twisted pair

Television is watched for hours with a typical program being at least one half hour in duration. Most telephone calls are short, usually a few minutes in length. The technical term for the length of a communication connection is holding time. The same television signals are sent to everyone; television is a broadcast medium. A telephone signal is connected to only one party; telephony is a switched service. Everyone can watch television at the same time, and most people watch during the evening hours known as prime time. The telephone network is designed to share resources among many potential users with the knowledge that only a small percentage will want to use the phone at the same time.

The only way in which CATV and telephony are similar at the local level is in the medium used to carry the signals. Today, CATV uses the copper wire of coaxial cable, and the telephone uses the copper wire of twisted pair. In the future, both will increasingly use optical fiber. However, as we have just seen, there are far more differences than just a similar transmission medium in our comparison of the two services.

There is a last way in which CATV and telephony differ. The CATV firm controls both the conduit—the cable—and the content—the TV programs sent over the cable. The telephone company owns and provides a switched, two-way conduit for communication but does not control, nor provide, the content of the communication. Indeed, the local telephone company is a monopoly, but the CATV firm is an even bigger monopoly. An even bigger monopoly could occur if the local telephone companies achieve domination over the provision of cable television while maintaining their control over the provision of local telephone service.

The conclusion that I make is that the two services are so different that separate networks make good engineering sense. However, the transmission path might be shared in a common cable containing the twisted pairs of telephony and the coaxial cable of CATV or whatever the separate transmission media of tomorrow may be.

FIBER TO THE HOME

A few years ago, much media attention was attracted by the tremendous capacity of optical fiber for delivering a wide variety of services, all in digital form. Voice, data, and video would be delivered to the home over the single, broadband medium of fiber. Thus, the concept of fiber to the home was invented, although it was simply a repeat of the promises of yesteryear's broadband medium, namely, coaxial cable. A number of technical problems were discovered with bringing fiber directly to each and every home, and as a result, fiber has shrunk back away from the home. Fiber to the curb and then fiber to the neighborhood were the result. The next sections discuss the technological challenges and uncertainties of fiber in the local loop.

THE DC BATTERY PROBLEM

I recall attending a conference about six years ago at which a panel of experts was discussing the prospects for fiber to the home, or its unpronounceable abbreviation FTTH. The usual hype and futuristic promises from a convergence of telephone service and television on a single broadband medium were being presented. A person from the audience dropped the clinker by asking how the direct current needed to power the phones in the home would be obtained, since optical fiber does not conduct electricity. The look of outright surprise and puzzlement by most of the panelists said it all—they had forgotten this little problem.

Telephones require direct current (DC) to work. The direct current is created at the central office and is conducted over the twisted pairs of copper wire to homes and offices. In an emergency, large batteries at the central office take over and are able to power the telephone system for hours, until a diesel generator at the central office is switched into operation. The reliability of this power system is quite high, and we all can remember storms and earthquakes when the lights went out but the telephone still worked.

There are technological solutions to the DC power problem. One obvious solution is to generate the dc at each home, or even at each telephone, by converting the 110-volt AC into DC, like the power adapter used for your telephone answering machine. The problem then is that the telephone would not work if the AC were not available, such as in an emergency. A battery backup could be used in the basement of each home for such emergencies, but then appropriate AC would be needed to charge the battery. Furthermore, rechargeable batteries ultimately wear out and would need replacement, thereby creating the question of who would be responsible for their replacement.

Not only do telephones in the home need direct current, but the sophisticated electronics needed at each home to demultiplex television and telephone signals also require direct-current power.

There are solutions to the DC battery problem. A system being developed by AT&T sends direct current over coaxial cable along with TV and telephone signals.

The sophisticated electronics in many of the new systems need to be protected from environmental hazards, such as heat, cold, and moisture. A solution is to bury the equipment in an underground vault, but even then, some protection in the form of air conditioning might be needed. But another big question is how to power this equipment and to protect it from power outages. Normally, the equipment will be powered from the AC of the local power company. But in an outage, back-up batteries will be needed, along with some form of small diesel generator. This power problem is solvable, but to do so is expensive.

NETWORK ARCHITECTURES AND TOPOLOGIES

The telephone network uses centralized intelligence at the central office and robust, simple telephones connected to the central office by simple twisted pairs of copper wire. The CATV network is even simpler since the TV signals are sent in one direction down the coaxial cable and only need to be amplified along the way. A very inexpensive "tap" or connector is used to attach each home to the coaxial cable. The use of fiber for CATV avoids the cost of amplifiers, but the connection of each home to the fiber then becomes a very costly and complex affair. Similarly, extending multiplexing technology and intelligence from the central office for telephone service greatly increases cost and complexity of the overall system. How the network is configured to deliver telephone or CATV service is thus an important technological issue.

The telephone network is called a star network since each telephone is connected by its own dedicated twisted pair of wires to the central office. The CATV network is a tree network since the signals from the head end branch to each home. Computers are frequently connected to a single transmission medium in what is called a bus configuration. The topic of network configuration is known as network topology.

The tree topology used for CATV is very simple and inexpensive. TV programs are multiplexed together at the head end and are then all sent down the coaxial cable. Along the way, the multiplexed signal loses amplitude and needs to be amplified every few thousand feet. Newer CATV system use optical fiber along much of the backbone to avoid these costly and sometimes noise-prone amplifiers. Homes are passed by the coaxial cable and are connected to it by very inexpensive connectors, called taps, that cost only a few dollars. A convertor box in the home selects and if necessary unscrambles the selected TV signal.

The coaxial cable is a broadband medium and could carry analog or digital signals. Some CATV companies are considering the use of compressed digital video so that more programs could be delivered over their existing cable. Some telephone companies are proposing a complete rewiring of the United States to deliver television over a totally new architecture and topology. In New Jersey, Bell Atlantic was granted regulatory relief in return for its promise to rewire the state with new digital network architecture.

The question is how to marry the advantages of a tree topology for delivering TV to many homes with the star topology of individual telephone lines for each telephone subscriber. One answer is the multiplexing of TV signals with telephone signals on the same transmission medium. If everything is converted to digital, then time-division multiplexing is appropriate. However, the equipment needed to demultiplex and multiplex signals at the very high bit rates needed for hundreds of video and telephone signals is quite complex and costly.

HYBRID FIBER COAX

One solution is called a hybrid fiber coax (HFC) system, and AT&T is pursuing this kind of approach in its product offering to the telephone companies and is proposing to install the system for Pacific Telesis in California. In 1994, Pacific Telesis predicted it would have 10,000 or more homes connected to HFC by the end of 1995, a million homes connected by mid 1996, and 5.5 million homes connected by the end of 1999. These predictions have been overly optimistic, and by May 1996, Pacific Telesis had connected 1200 homes in San Jose, CA in a technology trial of video over HFC with no commercial charges. Telephone service over HFC will come later.

With the HFC system, shown in Fig. 5-1, TV and telephone signals are carried over optical fiber from the central office to fiber nodes located in the neighborhood of customer homes. One optical fiber carries about 80 conventional analog TV signals. Two optical fiber pairs (with one pair for redundancy backup) carries digital telephone, data, and video signals. The TV and telephone signals are sent to the homes from the fiber node over a single coaxial cable shared by a few hundred homes. As many as 500 homes can be served by a single fiber node using about five coaxial cable buses. The connection of many fiber nodes to a single central office is a star topology and the coaxial cable network to the homes is a bus topology. The overall system is therefore called a star/bus topology.

The HFC system is also being installed by Southern New England Telephone (SNET) in Connecticut. A few dozen homes are being tested in 1996 with about 1,000 to be connected by the end of 1996. SNET predicts that the entire state will be HFC by the year 2009. SNET has already applied for a CATV franchise for the entire state.

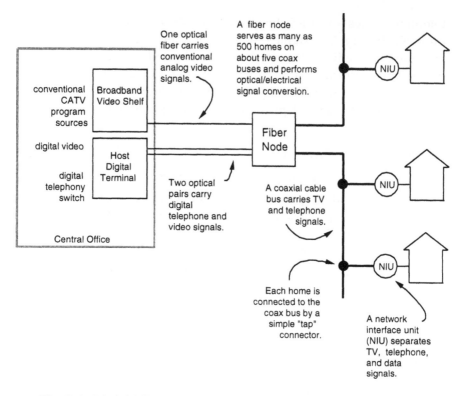

Fig. 5-1. A hybrid fiber/coax (HFC) system with a star-bus topology is being developed by AT&T for use by Pacific Telesis in California. PacTel expected to have 10,000 or more homes connected by the end of 1995.

FIBER TO THE CURB

A so-called fiber-to-the-curb (FTTC) system manufactured by Broadband Technologies, Inc. and being installed by Bell Atlantic in Dover Township, New Jersey uses fiber to deliver telephone service and video to homes. The system, shown in Fig. 5-2, provides 384 switched, compressed-digital TV channels, each at 6 Mbps using the MPEG-2 compression standard. The TV signals are switched and combined with the telephone signals by host digital terminals (HDTs) at the central office. An HDT serves about 300 subscribers. Pairs of optical fibers carry the multiplexed signals from the HDT at the central office to optical network units (ONUs) with each ONU serving about eight homes. Each home is connected to the pole-mounted ONU by its own cable

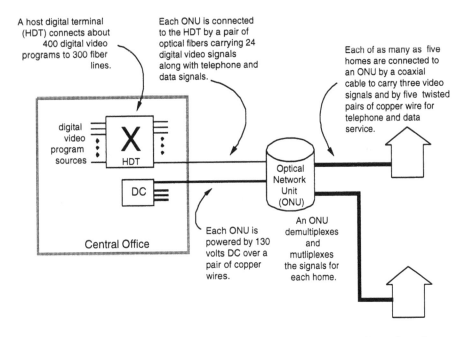

A host digital terminal (HDT) connects about 400 digital video programs to 300 fiber lines.

Each ONU is connected to the HDT by a pair of optical fibers carrying 24 digital video signals along with telephone and data signals.

Each of as many as five homes are connected to an ONU by a coaxial cable to carry three video signals and by five twisted pairs of copper wire for telephone and data service.

digital video program sources

X

HDT

DC

Central Office

Optical Network Unit (ONU)

Each ONU is powered by 130 volts DC over a pair of copper wires.

An ONU demultiplexes and mutliplexes the signals for each home.

Fig. 5-2. The double-star topology is being supplied by Broadband Technologies to Bell Atlantic for installation in New Jersey, with 10,000 homes expected to be connected by the end of 1995.

containing a coax for three digital video signals and five twisted pairs of copper wire for telephone and data service. Since all the fibers from the central office form a star topology and the cables from each ONU likewise form a star topology, the overall topology is therefore called a double star.

I visited a Bell Atlantic demonstration facility in the Summer of 1995 and was convinced that the technology seemed real and at hand. Bell Atlantic told me that they expected to have 10,000 homes connected to it by the end of 1995. By May 1996 about 5,000 homes were connected. There are problems though. Each ONU serves only eight homes and contains very costly technology, although neither Bell Atlantic nor Broadband Technologies would disclose the actual prices of the equipment. I doubt very much that this very costly system could ever compete economically with conventional CATV systems.

Each ONU is connected all the way back to the central office by a pair of optical fibers and by a pair of copper wires for DC power. With today's twisted pairs of copper wire, about 10,000 pairs are connected to a typical central-office switching machine. The Bell Atlantic/Broadband Technologies' approach simply replaces these twisted pairs of copper wire with pairs of optical fiber and heavy copper wires for power, although only about 1,000 pairs are needed. This is not an elegant solution. The key issue of most fiber-to-the-curb systems is their high cost.

These criticisms of fiber to the curb were confirmed on my tour of the Bell-Atlantic/Dover-Township installation in the Spring of 1996. I was shown the Bell Atlantic central office as part of the tour. My initial response was horror when I saw the large frame of equipment needed to support all the fiber pairs and to perform the video switching. This frame of equipment was about 7 feet high and nearly 50 feet long and handled 600 pairs of fiber serving about 4,000 homes. A full installation to serve 10,000 homes will require about three of these large frames of equipment. It all looked extremely complex and costly to me, but clearly the equipment does work and is real. Interestingly, like the Pacific Telesis system in San Jose, the Bell Atlantic system is being used solely for video—telephone service will come later.

ASYMMETRIC DIGITAL SUBSCRIBER LINE

Yet another approach is to use the existing twisted pairs of copper wire that are already in place. Twisted pair can carry as much as 6 Mbps over distances of a few miles, and most local loops are no more than a few miles from the central office. The approach, called asymmetric digital subscriber line (ADSL), sends a 6 Mbps digital TV signal to the home along with conventional two-way voice telephony over a single twisted pair. The approach is simple and utilizes the existing local loop, but costly video switching for each loop is needed at the central office. ASDL is probably an interim solution while we await improvements in technology and the results of today's trials.

TOPOLOGY UNCERTAINTIES

If you are like me, your mind becomes confused with all these different approaches and all their abbreviations and technical terminology. Which one is best? How much will they really cost? Are all the technical issues as resolved as the proponents of each one claim? Many people already complain about the price of today's CATV service. Given the sophistication and complexity of all these new approaches, will anyone be able to afford tomorrow's CATV service? The high expectations of the phone companies to have 10,000 homes connected by the end of 1995 have already been greatly curtailed.

I recall a study, done many years ago by Bell Labs when the world was analog, which concluded that the cost effective way to deliver both CATV and telephony was to use two separate networks: one using coaxial cable for TV and the other using twisted pairs of copper wire. There might however be cost savings if the coax and twisted pair shared the same cable, as in the AT&T system about to be installed for Pacific Telesis in California. From solely the provision of telephone service, Pacific Telesis expects savings of about $50 per year per subscriber. However, the technology is sophisticated and costly, and we will need to wait and see what technical problems might arise with the various systems that are being installed.

Today, the world is digital, and a fascination with video on demand has developed. What additional technological uncertainties are created when video on demand—rather than conventional cable television—is introduced?

VIDEO SERVERS

A server is a computer that is attached to a network and performs functions that collectively benefit the various users of the network. Video on demand (VOD) and video dial tone (VDT) will require access to many videos by many users on a network. These videos will be stored in digital form and accessed by a video server. Thus, a video server is a device, probably some form of computer, that will store the videos and enable access by many simultaneous users.

The performance requirements for a video server are quite staggering, even by today's standards for technology. If a single video is stored in compressed form at 4 Mbps, then a two hour video will require a total storage of roughly 30 Gbits, or about 4 Gigabytes (a byte equals 8 bits). To store 100 videos would then require 400 Gigabytes, or about one-half Terabyte; 1,000 videos would require 40 Terabytes. Today's hard disk technology stores in the order of a Gigabyte per platter, and a multi-platter hard drive could thereby store only about one video. One possible solution would be the use of a giant juke-box of video CDs, but other problems remain.

The problem of many simultaneous users who all want access to different portions of the video creates another challenge. Multiple heads could be used to allow access to each separate user, but physical constraints on the placement of the head mechanisms would probably limit the number of such users to about ten. What all this means is that the technology to allow hundreds and perhaps thousands of simultaneous users to have access to hundreds of videos simply does not exist today. There are ideas on paper about ways to develop such video servers, but commercial solutions are still far in the future.

Video servers capable of handling a few hundred users and containing perhaps a hundred videos were promised for some of the video-on-demand trials, but the technology simply was not available. The trials have had to be postponed or other alternatives used, such as a bay of a few hundred VCR machines with a robot arm, or a high-school student, to load the requested videos.

VIDEO SWITCHING

Even if the video server problem were to be solved, the next challenge would be how to switch each video to each user. The switching machines in today's telephone network are designed to switch speech signals at 64,000 bps, not video signals at 4,000,000 bps. Furthermore, the local switches are designed to handle only about 20 percent of customers simultaneously, which is why you have trouble obtaining dial tone during earthquakes and other emergencies when everyone tries to use the phone at the same time. Since nearly everyone watches TV during prime time, a video switching machine would need to handle 100 percent of the users simultaneously. Switching machines with the required capacity simply do not exist, and ideas on how to develop such machines are only now being designed on paper.

In some respects, the electromechanical switching machines of the past were more appropriate for the bandwidths needed for video. These machines simply created a physical path through the switching machine by closing electrical contacts, and electrical contacts can easily accommodate the high bandwidths and bit rates needed for video. Today's local, digital switching machines simply do not have the required capacity for video.

Another unresolved question is whether circuit switching or packet switching should be used. A number of technologists are promoting the use of packet switching in the form of the Asynchronous Transfer Mode, or ATM, which is a packet-switching technology allowing flexible packet lengths. Video has long holding times, unlike the short, bursty nature of most data traffic, and hence circuit switching might be more appropriate for video. This is just another technological uncertainty that must be resolved before the video switches required for video on demand are commercially available "off the shelf."

LANS AND ENGINEERING PRACTICE

Local area networks, or LANs for short, are data networks that interconnect personal computers and work stations so that software and resources can be shared and messages interchanged. Early LANs used coaxial cable since it was thought that the large bandwidth and capacity of coax was needed. Later LANs used optical fiber which offered even greater bandwidths.

An innovation in LAN technology was the use of twisted pairs of copper wire! It was discovered that twisted pair was sufficient in terms of bandwidth and capacity for the short distances between personal computers all located in the same building or office. Furthermore, twisted pair was far less expensive as a medium. It is interesting when the use of old-fashioned existing technology is considered innovative, and this lesson should remind us that good engineering examines all alternatives from a number of perspectives, such as installation, user needs, maintenance, and cost. Hype and the promotion of technology for its own sake have no place in sound engineering practice.

DATA TRAFFIC

The issue of network traffic has been controversial for decades. There are those who believe that data traffic will outstrip voice. Decades ago, such predictions were made and were incorrect. Today, such predictions are again being made and are used to justify broadband networks.

In an attempt to bring reality to the issue, I performed a study a few years ago to estimate traffic for various services by quantifying each service in terms of total number of bits per year. The single largest yearly service was local and toll telephone traffic which in 1988 amounted to $14{,}000 \times 10^{15}$ bits for the United States. To estimate what e-mail and data traffic might be, I assumed that all 120 million workers in the United States would access or create 200 screens of information per work hour. This came to only 300×10^{15} bits, much less than the traffic carried over the telephone network.

Some proponents of broadband data communication criticized my findings and claimed that transfers of large data files, containing images and vast amounts of other data, between personal computers would create tremendous amounts of traffic. In response to their criticism, I estimated this traffic by assuming that 20 million personal computers would each transfer 20 Mbytes per day. This came to a yearly traffic of $1,200 \times 10^{15}$ bits, still much less than the telephone traffic.

According to my estimates, the only kind of traffic that would surpass telephone traffic would be switched video, such as the videophone or video dial tone. Facsimile has had great growth in the last ten years, but it is carried over the telephone network along with voice signals. Much late night international traffic carried over the telephone network is most likely facsimile. In a later chapter, we will discuss the Internet and the packet-switched data signals that it carries. Such data traffic is minuscule, though, compared with the traffic carried over the switched telephone network.

DOWNLOADING

Many years ago while I was at AT&T, a colleague was squirreled away in his office working diligently on his new idea to download digital audio during evening hours over phone lines. I performed a back-of-the-envelope calculation and concluded that it would take over 40 hours to download one hour of CD quality music at 30 kbps, a fairly fast rate for a regular telephone line.

Clearly, my colleague's idea was not practical. Yet the idea of downloading video, which would require more capacity and time than audio, has surfaced. What continues to be forgotten is that the least costly way to deliver vast amounts of information is physically, such as a compact disc or a video tape or a newspaper. A network with electronic delivery is required only for timely information, and broadcasting the information is far less costly than a switched two-way network.

ANYTHING'S POSSIBLE ON PAPER

There will be those who will claim that they have solutions to all the technological uncertainties that I have listed in this chapter. But their solutions will frequently be ideas that thus far are only on paper. Or if their ideas have made it to actual hardware, the hardware will still be in the laboratory and not yet available"off the shelf" for installation in real situations.

Indeed, nearly anything is possible and solvable technologically, but ultimately there is a cost associated with technology. For example, the rewiring of the Nation proposed by many of the Baby Bells to bring us more TV programs utilizes very costly technology and does little more than conventional CATV. Will the service be affordable to consumers? Who will finance the required investment? These questions bring us to the next chapter in which we examine the financial viability of the superhighway.

READINGS

Kyees, Philip J., Ronald C. McConnell, and Kamran Sistanizadeh, "ADSL: A New Twisted-Pair Access to the Information Highway," *IEEE Communications Magazine*, April 1995, pp. 52-59.

Noll, A. Michael, "Rethinking the Digital Mystique," *Telecommunications*, Vol. 30, No. 2, February 1996, p. 43.

Reed, David P. and Marvin A. Sirbu, "An Engineering Cost and Policy Analysis of Introducing Fiber into the Residential Subscriber Loop," *Broadband Networks*, ed. Martin C. J. Elton , North-Holland: New York, 1991, pp. 89-134.

Schwartz, Evan I., "Demanding Task: Video on Demand," *The New York Times*, January 23, 1994, Business Section, p. 14.

Young, Garvin & Don Clarke, "Video Delivery over Copper Pair and Its Role Relative to Fiber," *SMPTE Journal*, June 1966, pp. 351-356.

"Tomorrow's network, tomorrow," *The Economist*, April 20, 1996, pp. 57-58.

Chapter 6

FINANCIAL VIABILITY

Ultimately, someone must make a buck for any product or service to be successful in the real world. The financial investment required to create the infrastructure needed for the superhighway is quite large. The investors would want adequate returns on this investment, which means that consumers would need to pay the amounts required to generate this adequate return on investment. We must therefore examine the financial viability of the superhighway, giving adequate attention at the broadest level to the overall financial structure of the communication industry and some of its major players. In this chapter we will also assess the financial viability of some existing and proposed new superhighway ventures and discuss a number of financial issues affecting telecommunication broadly. Consumer willingness to pay will be covered in a later chapter on consumer need.

COMMUNICATION INDUSTRY REVENUE

The communication industry is huge: about seven percent of the Gross Domestic Product. The following table gives industry revenue for 1993 assembled from data reported in the *Statistical Abstract of the United States: 1995*. Telecommunication is the largest segment of the communication industry and is nearly twice as large as any other segment.

The $48 billion in revenue generated by motion pictures includes $6 billion from movie theaters and $7 billion from tape rentals, indicating the extent to which tape rentals have grown in a short few years. The sale of television equipment was $15 billion. Half of the $27 billion from CATV came from basic service; pay-per-view and premium generated only about $5 billion. The newspapers, periodicals, and books produced by the publishing industry rival the revenues of cinema and television. Radio broadcasting is very much alive and generated $8 billion of the audio industry revenue. Most of the revenue of the personal computer industry came from the sale of equipment. Physical delivery generated as much revenue as the entire publishing industry.

Segment	1993 Revenue
TELECOMMUNICATION	**$221 billion**
• Telephone Service	$178 billion
• Equipment	43 billion
CINEMA & TELEVISION	**$112 billion**
• Motion Pictures	$48 billion
• Television	37 billion
• CATV	27 billion
PUBLISHING	**$78 billion**
• newspapers	$34 billion
• periodicals	22 billion
• books	22 billion
AUDIO	**$28 billion**
• radio stations	$8 billion
• recorded media	10 billion
• equipment (home & auto)	10 billion
PERSONAL COMPUTER	**$37 billion**
• Hardware	$30 billion
• Software	7 billion
PHYSICAL DELIVERY	**$77 billion**
• U. S. Postal Service	$48 billion
• United Parcel Service	20 billion
• Federal Express	9 billion

Some interesting observations and conclusions follow from the above table. The total revenue of the CATV industry is considerably lower than that of the telephone industry, and the revenue generated from premium CATV is comparatively small. Why then is the telephone industry interested in CATV? It would make much more sense for the CATV industry to provide telephone service, since telephone service generates considerable revenue and thus is a much larger target.

The conventional accepted wisdom is that the revenue is in content and not in the conduit. Yet, the provision of telephone service, which generates no revenue from the provision of content, generated over half as much revenue as content-based cinema and television. Perhaps what this really means is that the provision of content is central to entertainment media, such as motion pictures, television, and audio, but not necessarily to other segments of the communication industry.

LESSONS FROM THE FINANCIALS

The financial performance of a company is summarized in its consolidated statement of operations. Revenues and various expenses are deducted to determine the final profit, as shown in the table below. Deductions are shown in parentheses. Cash flow is sometimes called

CONSOLIDATED STATEMENT OF OPERATIONS	
ITEM	**MARGINS** as percentage of operating revenues
Operating Revenues	
(Operating Expenses)	
CASH FLOW or EBITDA	CASH FLOW MARGIN
(Depreciation & Amortization)	
OPERATING INCOME	OPERATING MARGIN
(Interest)	
Other Income & Expenses	
PROFIT BEFORE TAXES	PRE-TAX PROFIT MARGIN
(Taxes)	
Other Items	
NET INCOME	PROFIT MARGIN

EBITDA, for Earnings Before Interest, Taxes, Depreciation, and Amortization; it equals revenues less direct expenses.

Operating income equals EBITDA less depreciation and amortization. Pre-tax income equals operating income less interest and any miscellaneous revenues and expenses. These items are frequently expressed as percentages, or margins, on revenue.

We will now compare the financials for a selection of media and telecommunication firms, including AT&T, Bell Atlantic, TCI, Time Warner, and TCA. AT&T is the former owner of the old Bell System and today provides long distance service, mobile telephone service, and manufactures various telecommunication and computer equipment. Bell Atlantic is one of the so-called Baby Bells. Tele-Communications, Inc. (TCI) is the largest MSO in the United States. At the end of 1995, TCI supplied cable television to 12.5 million subscribers. Time Warner Inc. is a large multi-media company publishing various magazines (such as *Time* and *People*) and recorded music. Time Warner Entertainment Company, L.P. (TWE) was spun off from Time Warner, although Time Warner still owns 63 percent, and provides filmed entertainment (Warner Bothers), television programming (HBO), & CATV. TCA Cable TV, Inc. is the 17th largest multiple system operator (MSO), with 57 CATV systems and 580,000 subscribers. TCA provides CATV service mostly to areas that are not covered by conventional broadcast television, and thus TCA's markets have a higher penetration of CATV subscribers than in those markets served by over-the-air broadcast signals.

The table on the next page summaries the 1995 financial performance in terms of margins on revenue for selected media and telecommunication firms ranked by profitability. More detailed financials for these and other telecommunication firms are given in appendix B to this book.

The bottom line for these companies is that TCA and Bell Atlantic are quite profitable with comparable profit margins of 16% and 14% respectively. TCI operated at a net profit margin of only 1% in 1994, after having net losses in 1993, 1992, 1991, and 1990, although the size of these losses has steadily decreased. The reason for this low profitability is that in 1994 TCI paid nearly 16% of its revenues as interest versus TCA and Bell Atlantic which paid only 7% and 4%, respectively. Clearly, TCI is a debt ridden company without the profits to reduce its heavy debt. What is truly baffling is that in 1994 Bell Atlantic attempted to acquire TCI. The acquisition of TCI's debt surely would have reduced the financial success of Bell Atlantic. The only apparent rationale for the ill-fated merger was Bell Atlantic's belief in the hype of the highway of dreams. Anyone who understood the negative implications of the financial aspects of this merger sighed with relief when the deal fell through.

	1995 FINANCIAL PERFORMANCE			
	CASH FLOW MARGIN	OPER-ATING MARGIN	PRE-TAX PROFIT MAR-GIN	PROFIT MARGIN
TCA Cable TV, Inc.	50%	35%	28%	16%
Bell Atlantic	42%	23%	22%	14%
AT&T (1994)	16%	11%	10%	6%
Time Warner	15%	8%	-0.5%	-3%
Time Warner Entertainment	21%	10%	2%	1%
Tele-Communications, Inc.	37%	16%	3%	1%

The above table shows how different some of these firms are in terms of their financial performance. Clearly, being a telephone company or a small CATV company can be very profitable and successful. Bell Atlantic and TCA are money-making machines, with high cash flows, low interest, and healthy profit margins. However, being a large CATV company or a multimedia company is not necessarily very profitable. However, an extremely large CATV company (such as TCI) can generate considerable cash flow, although large payments of interest to service large amounts of debt greatly shrink the income. The multimedia company Time Warner clearly is having financial difficulties. TWE was spun off a few years ago to attract new partners, such as US WEST, Inc. (25.5% ownership), ITOCHU Corporation (5.6%), and Toshiba Corporation (5.6%). However, TWE is not performing that well either. The problem at both Time Warners seems to be very high operating expenses along with large interest payments.

In 1993, Bell Atlantic and TCI had quite similar cash flows and operating margins, but then the similarity suddenly stopped. The pre-tax profit margins were very different; TCI pays considerable interest on much debt. TCI thus is a highly leveraged company. Bell Atlantic was wise to cancel its planned acquisition of TCI, since Bell Atlantic would have dramatically and negatively changed its financial profile with this acquisition.

The table further educates us about the pitfalls of manufacturing by showing the financial performance for AT&T. Since AT&T had pre-tax business-restructuring and other charges of $7.8 billion for 1995, the financial performance for 1994 is used as being more typical. AT&T's cash flow margin for 1994 was considerably lower than the cash flow margins for Bell Atlantic and the two CATV firms analyzed previously. The other margins are roughly half those for Bell Atlantic and TCA. The reason for this is that AT&T is a diversified company with a large manufacturing component, and manufacturing traditionally has very tight margins. Although AT&T's long distance business probably has very good margins, its cellular business obtained through the acquisition of McCaw Communications has most likely not been a strong positive factor on its overall margins thus far.

Given this data about the financial performance of their former owner, one wonders why the Baby Bells ever wanted to be allowed into manufacturing, unless they are looking for ways to lose money. Many of the Baby Bell new ventures have not been great money makers.

Another measure of financial performance is return on investment, or ROI. The return is profit either before or after taxes. The investment is assets minus liabilities and is also known as shareowners' equity. For 1994, the return on investment for AT&T was 26%; for Pacific Telesis, 22%; and for TCI, 1.6%. One problem with return on investment is that it can be inflated if a company, like a telephone company, owns much plant and depreciates it slowly thereby increasing the value of its assets. Furthermore, return on investment will be deflated if a company has much debt, like TCI. Thus, return on investment can be a useful measure for comparing companies in identical businesses, but return on investment is not illuminating when used as a measure of comparison for companies in different businesses. Return on investment has been used as a basis for regulation of telephone companies, but has been replaced by caps on the prices charged to customers, subject to incentives for improvements in productivity.

HOW'S THE OLD BELL FAMILY DOING?

Appendix B gives the 1995 financial performance in terms of margins on revenue for each Baby Bell and also for AT&T for 1994. As before, cash flow equals revenues less direct expenses; operating income equals cash flow less depreciation and amortization; and pre-tax income equals operating income less interest and any miscellaneous revenues and expenses.

The Baby Bells main source of income is their monopoly provision of local telephone service, and one would expect their performance to be similar. However, NYNEX's performance is clearly below average. AT&T has a large manufacturing business in addition to its provision of long-distance service, and all aspects of AT&T's business are subject to intense competition. Not surprisingly, AT&T's 1994 performance does not even come close to the performance of an average Baby Bell.

For comparison purposes, the table of 1995 financial performance includes GTE, now the largest regional operating company. GTE had revenues in 1995 of nearly $20 billion making it larger than the closest Baby Bell, namely Bell South, which had 1995 revenues of nearly $18 billion. About one-fourth of GTE's business is in providing local telephone service internationally, which makes GTE somewhat different than the typical Baby Bell.

Regulators have traditionally regulated the return on investment of local telephone companies. However, the level of investment depends on depreciation and other discretionary factors. The financial community looks more strongly at margins on revenues, which is why the preceding analysis also uses such margins to examine financial performance.

A measure of performance traditionally used in the telephone business is productivity, which is measured as the number of access lines per telephone employee. The table on the next page gives this measure as reported by each of the seven Baby Bells, although I worry that the criteria to identify and count telephone-related employees might not be identical for each. However, I have performed my own estimates of productivity using the total number of employees reduced by the proportion of network revenue to total revenue, and my rankings are nearly identical to the Baby-Bell reported figures.

BABY BELL PRODUCTIVITY	
	Access Lines Per Telephone Employee
Ameritech	373
Bell Atlantic	326
Bell South	308
NYNEX	287
Pacific Telesis	347
SBC Corporation	294
US WEST Communications	310

One of the Baby Bells with the highest productivity is Pacific Telesis. This is the result of two factors: an intense campaign to reduce its number of employees and its divestiture a few years ago of non-telephone related business. Ameritech has the highest productivity, perhaps because of the seven Baby Bells, Ameritech has been most focused on its core telephone business. NYNEX's poor financial performance is clearly related to its low productivity and indicates that NYNEX still has too many employees. The significant difference in productivity between Pacific and SBC, and between Bell Atlantic and NYNEX, raises questions about the outcome of these two mergers.

Executive pay has become a touchy topic recently. Although NYNEX's financial performance in 1994 was far below the other Baby Bells, the NYNEX chairman received a salary and bonus of $1,685,000 which was exceeded only by SBC Corporation's chairman who received $1,952,000. In 1994, Pacific Telesis showed excellent financial performance and the best productivity of the Baby Bells, and its chairman received $933,000, which was the lowest pay of the seven Baby Bells. The Baby Bells make most of their profits from their provision of basic telephone service, which is a utility business. One would perhaps expect that their chief executives would receive pay similar to that of other utility companies. Such large power utilities as Pacific Gas and Electric, Con Ed, and Public Service Electric and Gas each paid their chairmen less than $1 million in 1994. Why then are most of the Baby

Bells paying their top executives such large amounts?

What is most illuminating about these financials (see Appendix B for details) is that in 1995 the seven Baby Bells had aggregate profits of $11.8 billion on an aggregate revenue of $91.6 billion. The Baby Bells generously pay a large percentage of their profits as dividends to their shareowners, actually as much as 80 percent for some. If the Baby Bells were to reduce the dividend pay-out ratio to 50 percent, they collectively could have nearly $6 billion to invest. Very little would be beyond such tremendous collective financial power. But their track record at investing in new ventures has not been very successful, justifying the term "Baby Bells in the woods."

WHERE ARE THE PROFITS?

In early 1995, I attended a financial conference at which Sprint presented very interesting data about the 1994 operating margins for its various businesses. Cellular and wireless had an operating margin of 14%; long distance was only 9%; and local service was 23%. The Sprint operating margin for local service is very similar to the operating margins for the Baby Bells and GTE.

The surprise in the Sprint data is that the provision of local telephone service is such a great financial success—more so than cellular or long distance for Sprint. It also becomes clearer why GTE sold Sprint and abandoned the long distance business. Perhaps the real cash cow is the provision of local telephone service. This conclusion is enough to create curiosity about how well the various new business ventures of the Baby Bells are performing financially.

BABY BELL NEW VENTURES

Pacific Telesis' 1992 Annual Report was most revealing of the profitability of new ventures. It presented the financial and operating results for Pacific Telesis' new ventures for 1990 through 1992. Pacific Telesis International (PTI) operated cellular and paging systems in Germany, Portugal, Spain, Thailand, and Japan. PTI had a net loss for all three years with an accumulated loss of $93 million on accumulated revenues of only $78 million. PacTel Teletrac, a joint venture providing vehicle location service, also had net losses for the three years with an accumulated net loss of $78 million. Only PacTel Cellular and PacTel

Paging had net profits for the three years: a total accumulated net income of $312 million on accumulated revenues of $2,106 million—a hefty profit margin of 15% over the three years. However, the analysis for 1992 presents a negative picture: $82 million net income for PacTel Cellular; $16 million net income for PacTel Paging; a net loss of $33 million for PacTel Teletrac; and a net loss of $46 million for PTI. The overall net income for 1992 for these ventures was only $19 million. In 1994, Pacific Telesis wisely divested all its paging and cellular ventures into a new company called AirTouch Communications.

US WEST, through TeleWest Communications, offers combined telephone/CATV service in the United Kingdom. The US WEST investment in TeleWest since 1989 has been $420 million, but in 1994 the UK telephone/CATV business had a loss of £58 million on revenues of £72 million, which is an increase over the losses of the previous years, although revenues have been increasing also. US WEST's cellular subsidiary had a cash flow margin of 28% for 1994, which is far below the overall cash flow margin of 42% for US WEST as a whole. What all this means is that many of US WEST's ventures are either outright losers or, at best, drains on the overall profitability of the company.

INFRASTRUCTURE INVESTMENT

A few years ago, studies mostly financed by the Baby Bells concluded that the telecommunication infrastructure in the United States had fallen behind other industrialized countries. Anyone who has visited other countries knows that we in the United States have one of the best, if not the best, telecommunication systems. Clearly, the studies were flawed and were little more than a thinly disguised attempt by the Baby Bells to justify less regulation to allow them to rewire the country. The fact is that telecommunication companies in the United States already invest heavily in improving and maintaining their networks.

CATV companies and phone companies need to spend constantly in upgrading and maintaining their network infrastructures, and in this way, both industries are similar. For example, in 1993 Bell Atlantic spent $2,519 million, or 19 percent of its revenues, in upgrading and maintaining its telephone network. Also in 1993, TCI spent $947 million, or about 23 percent of revenues, in upgrading and maintaining its CATV network. The investment needed for the video superhighway, however, would dwarf these figures.

The estimates of the investment needed to rewire each home for the video superhighway range from $500 to $2,000. Let us assume an in-between estimate of $1,000 per home, which is probably on the low side. Rewiring the 100 million households in the United States would then require an investment of $100 billion, which is nearly half the value of the present local telephone system, which took over 100 years to create. Assuming an interest rate of 15 percent and a 20-year equipment life span, this level of investment is equivalent to $16 billion per year. Even the Baby Bells do not have these kinds of resources.

ACCESS CHARGES: THE FINANCIAL GLUE OF THE TELECOMMUNICATION INDUSTRY

Before the Bell breakup, whenever you made a long distance call about 60 to 70 percent of the charges were returned to the local telephone company as a form of subsidy to keep local rates low and to make telephone service affordable. After the Bell breakup, this subsidy has continued but is now justified as a reimbursement to the local company for the use of its local facilities to complete the long distance call. The reimbursement is known as an access charge. Long distance companies pay the local phone companies at each end of a call for access. You yourself also pay the local company an FCC-mandated flat rate of $3.50 per month for access. This user access charge is really an increase in local rates that is disguised by the local phone companies as some form of mysterious, FCC imposed charge. My guess is that some people believe that the fee goes to the FCC rather than directly to the local telephone company. In a way, the access charges paid to the local companies are like a divorce settlement.

The access charges paid by long distance companies to the local phone companies are quite large and are a form of glue that keeps the old Bell System together financially. In 1993, 24 percent of Bell Atlantic's total revenues came from access charges, thereby making long distance companies Bell Atlantic's biggest customer. In 1994, AT&T paid 41 percent of its revenues from the provision of long distance service in access charges and other interconnection charges. Thus, the local telephone companies, though currently forbidden from providing long-distance service, receive about 40 percent of long distance revenues because of access charges. These charges have been decreasing though. In 1984, AT&T paid 57 percent of its long distance revenues in access charges. Over this eleven-year period, AT&T's long distance revenues increased only 20 percent. In 1994, 25 percent of the total revenues of the Baby bells came from the collection of access charges.

Indeed, local phone companies need to be reimbursed for local access, but, making an honest determination of what these charges should be, is another matter. The local companies have a monopoly on local access and are a bottleneck to local access. The AT&T acquisition of McCaw's cellular business gives AT&T a detour around the local companies. One can imagine AT&T giving a lower long distance rate to its McCaw customers, who use the cellular connection as a direct route to AT&T's long distance network, thereby bypassing the local phone company entirely and saving on any access charges. One phone company response to this bypass is to increase local rates to its remaining customers, so that the end result is revenue neutral to the local phone company. Another response by the local phone companies is to clamor to be allowed into long distance service itself. The Telecommunications Act of 1996 indeed allows the local telephone companies into long distance.

As the largest provider of long distance service in the United States, AT&T, through its payment of access charges to the local phone companies, is their single largest customer. In return, the local phone companies purchase AT&T switching machines and are as a group AT&T's largest customer. AT&T and the Baby Bells are still intimately intertwined financially, although like parties in a divorce, they are battling more and more. AT&T will spin off its manufacturing into a new operation, called Lucent Technologies. This will eliminate the customer/supplier relationship between the Baby Bells and AT&T in the purchase and sale of switching machines, but now AT&T and the Baby Bells will battle even more fiercely over long distance.

INTERNATIONAL CALLING

When you make an international telephone call from the United States, you are most likely subsidizing the telephone system or government of the country you have called. This is because rates for calling from the United States to most countries are much lower than the reverse rates for calling to the United States. In fact, in 1993, over twice as many calls are made internationally from the United States than to the United States. This in itself is not a problem, but the way the costs of an international call are shared creates an inequity. The process of computing these shared costs is called settlements. The settlement rate is shared equally between the two countries for an international call, and if traffic were equal, there would be no net flow of money. However, given the inequity in origination of international calls and also the artificially too high settlement rate, a large imbalance occurs. In 1993, this imbalance was $4 billion from the United States and is growing.

Many countries view this imbalance as a cash cow generating funds that can be used by the government for any purpose and thus are very reluctant to renegotiate the settlement rate to more accurately reflect the real costs of terminating a call in their country. In many ways, the settlement rate is like the access charges charged to long distance companies.

USAGE SENSITIVE PRICING

Local telephone calls are usually free within a basic flat rate. This creates false economics when it comes to data traffic for which long connect times are involved. Long distance companies pay about 3 cents per minute to both the originating and terminating local phone companies for local access. If all local calls were charged similar rates, an hour of local connection would be priced at about $2.

We shall see in a later chapter how flat rates for access to the Internet create a subsidy to high volume users. The Internet and its precursor, the ARPANET, were subsidized by the federal government for over two decades. Commercial public data networks, which can not afford to subsidize users, charge for both connect time and the amount of data.

PRODIGY: A FINANCIAL ANALYSIS

IBM Corporation and Sears, Roebuck and Co. launched an electronic information service called PRODIGY® a few years ago. The service was reported to have cost between $500 million to $1 billion to develop. It was further reported that in early 1995 the service had about 1.3 million paying users and about 750 employees. This is enough data to perform a rough financial analysis of Prodigy.

The initial investment needs to be recovered with an adequate return. Assume that the initial investment was an average of the upper and lower estimates, or $750 million. Assume further that this initial investment should be recovered over a period of 15 years with a return of 15 percent. This would require $128 million yearly. A recovery over a shorter period of 10 years at a higher rate of 20 percent would require $179 million yearly. This tells me that a business person would want Prodigy to generate profits in the order of $150 million yearly to recoup the original investment and generate a reasonable return.

Prodigy's income stream comes mostly from the monthly rates paid by its subscribers. Assume that each subscriber generates $13 monthly in income from all sources. With 1.3 million subscribers, this amounts to about $200 million yearly. Assume that each Prodigy employee has loaded costs of $130,000 a year. The 750 employees thus create an expense of about $100 million. Additional expenses would be incurred for advertising, billing, maintenance, rent, and operations. Assume that these other expenses double the employee expense thereby giving total operating expenses of $200 million. These expenses would exactly balance operating revenues. Prodigy would thus not generate profits on an operating basis, and most certainly would not be generating enough profits to pay back the original investment. Prodigy clearly became a financial disaster for IBM and Sears.

IBM and Sears must have had much faith and hope for the future to continue in the venture, although once the initial investment was made there seemed little that could be done to recoup it. It seemed that as long as Prodigy was not losing money on an operating basis, nothing much was lost by continuing a little longer. But faith frequently loses to reality, and IBM and Sears looked to dump Prodigy. In May 1996, Prodigy's management, with financial backing from International Wireless Inc., bought Prodigy for a deal reported to be worth nearly $200 million. Given the fast growth of Internet suppliers and decreasing subscriber base to Prodigy, one wonders whether the new Prodigy owners will have much success. IBM and Sears, however, lost their shirts, and clearly a few $100 million is better than nothing.

THE BELL ATLANTIC/TCI MERGER: A FINANCIAL ANALYSIS

In early 1994, the proposed acquisition of the CATV-giant TCI by Bell Atlantic came unraveled. The purchase price was never very clear but seemed to be at least $2,000 for each of TCI's CATV subscribers. Assume that this $2,000 is an initial investment which must then be repaid and must also generate an adequate return. For a high-risk investment such as this, a return of 20 percent over a period of 15 years would seem appropriate. Each CATV subscriber would need to generate profits of $35 per month to create such a return. The average CATV subscriber bill in 1994 was $33 per month. Since there are such costs as content, billing, and maintenance, it seems obvious that the type of return just mentioned could only be generated by a substantial increase in CATV billing to perhaps $70 per month. Such a huge increase clearly would be impractical.

The Bell Atlantic merger with TCI made no sense financially, as we already saw earlier, based on the differences in overall financial structure of the two companies.

BANDWIDTH COSTS MONEY

In a previous chapter on technology we saw that with optical fiber, bandwidth is virtually unlimited. This tremendous capacity has lead some experts to conclude that bandwidth will be essentially free in the coming world of the superhighway. In my view, this conclusion is flawed.

Although tremendous amounts of bandwidth are now available in most telecommunication networks, bandwidth is not free. A large investment in technology has been required to install the transmission media that contain such bandwidths. Bandwidth in transmission media can be used to carry a variety of different services, with different services requiring different bandwidths, or capacities. How these different services are priced relative to each other can not be ignored. In this respect, bandwidth will never be free.

Consider a transmission medium that can carry both digital telephone signals at 64 kbps per channel and digital TV signals at 1.5 Mbps. What this means is that 24 telephone channels could be placed in the capacity of one TV signal. If the common carrier charges considerably less than 24 times the telephone service for the TV service, an entrepreneur could purchase capacity at the TV rate and then provide telephone service at a lower price. This would undercut the telephone service provided by the common carrier who would then need to increase the TV rates in an attempt to achieve equilibrium. Bandwidth can not be divided without consideration to its economic value.

TV VIEWER MARKET SHARE

As a market is divided again and again, the size of each segment is greatly reduced. For example, with 500-channel television, the average number of viewers for each channel would be 1/500th the total number of viewers. If each individual program cost $500 thousand to produce and distribute, then all 500 channels would need to generate $250 million per hour in revenues. Assuming a viewing day of 10 hours, this amounts to a yearly revenue requirement of one trillion dollars (or $1,000 billion). Clearly this makes no sense. The basic economic fact is that as the market is divided more and more, there is less and less money available for quality programming.

One possible outcome is that quality programming would disappear. Another outcome is that the same program would be sent over a number of channels with each channel starting the program delayed 10 minutes or so. This would allow the viewer to watch the program whenever the viewer wanted to rather than at a fixed schedule. Such a scheme has been called "near video on demand."

TECHNOLOGICAL INNOVATION AND PRICE

Indeed, technological innovation has greatly improved our lives. Technological innovation in electronics and in telecommunication has not only resulted in improved performance but also in much lower prices to consumers. I recall as a child that my parents purchased their first TV set on time since it was so very costly. Today's TV set is priced much lower even without correcting for inflation, and offers vastly improved performance, reliability, and quality. Video on demand might eliminate the need for a trip to the video store to rent a video, but the real question is how much this convenience is worth to the consumer. It might be that the trip to the video store is an excuse to leave the home and is not the inconvenience that is assumed. But, if video on demand is an improvement, perhaps it should follow the usual course of technological innovation in telecommunication, and cost less than renting a video from the video store. However, the promoters of video on demand assume that consumers will pay a premium. This assumption might be false.

In the present chapter we have seen that there are many financial uncertainties and difficult financial issues that make the future of the information superhighway very unclear. However, in the end these financial issues become intertwined with consumers and their needs. This then brings us to a host of questions about consumer acceptance and consumer need for the various services that would be delivered over the superhighway and about whether consumers will be willing to pay for them. These are the topics of the next chapter.

READINGS

Anthony, Robert N., *Management Accounting (Fourth Edition)*, Richard D. Irwin, Inc.: Homewood, IL, 1970, pp. 293-313.

Vogel, Harold L., *Entertainment Industry Economics (Third Edition)*, Cambridge University Press: New York, 1994.

"They lose their shirt," *The Economist*, November 19, 1994.

Chapter 7

CONSUMER NEED

Technology is only successful if it satisfies the actual needs of real consumers. Decades ago, AT&T discovered that all the promotion in the world could not convince most people that they needed a picturephone. Technology offers—but consumers decide.

The present chapter discusses the consumer considerations and reactions that will shape the future of the telecommunication superhighway, giving examples from today. One example is caller-ID and the surprise of negative consumer reactions to it. The chapter ends with some general background on the various means for determining consumer needs and their reactions to new products and services.

CONSUMER NEEDS

We already know that people have many needs, some more basic than others. People need shelter, food, and clothing. Although a tent of tree branches might have been adequate for our ancestors, many of us need large, fancy houses in specific neighborhoods, with rooms that might be used only an hour a day for some specific purpose. Similarly, while an animal skin might have been adequate clothing for our ancestors, an entire fashion industry has developed to convince us that we need this season's newest fashion rage to feel pleased with ourselves. So, the means for fulfilling these three simple basic needs have evolved into today's large industries that convince and entice us to purchase their products and services.

People have other needs too. We need security and means of

transport also. While we each might have carried a sword or other form of weapon not that long ago, such individualized security measures are today frowned upon and instead society hires police to secure our property and selves. The horse and buggy has become the automobile, but the automobile has evolved far beyond simple basic transportation to encompass style and status. We need to be entertained. Songs and stories around the fire that created a feeling of shared community have become variety shows and soap operas viewed on the TV set.

These general needs are interesting to list and think about, but they do not really help us understand new opportunities and the future. We need to be more specific. As an example, consider our need for privacy—the need to protect the inner self.

CALLER-ID

Caller-ID is a service that allows you to know the telephone number of the caller before you answer the telephone. Caller-ID is based on technology known as signalling system 7 (or SS7 for short) that sends the telephone number of the calling party over the telephone network. The telephone number is then sent down your telephone line as a short burst of data before you answer the phone.

The telephone company was convinced that many people would like to know who was calling before answering the phone. After all, who would open the front door without first knowing the identity of the person ringing the door bell. You can well imagine the surprise of the telephone company when people objected to the service, claiming that the privacy of their telephone number would be violated. In hindsight, this should not have been such a large surprise, since in many big cities large numbers of people pay extra to have unlisted telephone numbers. A few years ago, I did a study that also helps illuminate the issue of caller-ID from a consumer perspective.

My study questioned 43 graduate business students to determine their ratings of concern on a scale ranging from 0 to 5 (with 0 being "no concern" and 5 being "very concerned") for a variety of communication situations. "Not knowing who is calling before answering the telephone" received an average rating of only 0.9. "Your telephone number is displayed to the person you are calling" received a much higher rating of 2.5. What this means is that the solution—caller-ID—is of far greater concern than the problem it would solve, namely, not knowing who is calling before answering the telephone.

As a result of this consumer concern over caller-ID, some telephone companies have been required to give consumers an option to block the forwarding of their telephone number. Other telephone companies, feeling that administrating the blocking option was simply too much trouble, delayed offering caller-ID in their service areas. Today some telephone companies charge for blanket caller-ID blocking on all calls, while others offer blanket blocking for free. We are thus starting to find significant differences in the nature of telephone service in different parts of the United States.

INFORMATION SERENDIPITY

I read the newspaper mostly for enjoyment, not to find specific facts or articles. I really do not know what information I am looking for. I read the L. L. Bean and other catalogs similarly for fun without any specific product in mind. When I wander through the bookstore, I am likewise just browsing to see if anything might interest me. If something does, then I rapidly scan through a book to determine whether I would enjoy reading it. Much information browsing is of a serendipitous nature, and this makes it very difficult to create an electronic profile so that an information service on the superhighway could know your information needs in advance.

The capacity needed to enable me to browse electronically through the high-quality, color images on the pages of the L. L. Bean catalog would be astronomical and very costly, unless the government or somebody else subsidized the network, which indeed has been the case for the Internet. But even if the capacity were available and affordable, most people seem to prefer reading material on paper rather than on a computer screen.

Earlier, in the description of the technological Utopia, we read about home real estate shopping by computerized profiles. Such systems have been developed and have failed, again because most people are not able to articulate their specific needs. A good real estate agent knows how to observe the reactions of a client and how to ferret out specific, unspoken needs and lifestyles to match the client to a home.

There are times, however, when I actually look for specific information. Assume I know I want to buy a green widget in size 78. I now want to know who has the best price and the quickest availability.

An electronic service would be very helpful for this form of information and could save me much time and trouble, but it would have to be easy to use and not cost extra.

CONSUMER EXPENDITURES

The revenues that many promoters of the information superhighway expect to generate defy imagination. Realistically, consumers have finite incomes, although the information highway companies seem to have unlimited resources to spend in their quests for the Grail. A way to shed some light of realism on these extraordinary expectations is to examine how the average household currently spends its money each month. The following table presents an estimate of these monthly expenditures and is based on data reported publicly. Where indicated, the data is the number of households using each specific product or service, appropriately normalized.

MONTHLY EXPENDITURE (per household)	CATEGORY
$363	Food
$107	Gasoline (89% HHs with motor vehicles)
$62	Telephone Service (94% HHs subscribing)
$61	Video, Audio, & Computer Equipment, & Musical Instruments
$33	CATV Service (62% HHs subscribing)
$18	Books
$17	Magazines, Newspapers, & Sheet Music
$15	VCR Tape Rentals & Purchases (72% HHs with VCRs)
$5	Movie Theater Admission

The $33 per month spent on CATV looks like a tempting target, but only about two-thirds of TV households subscribe to CATV. It is reported that Bell Atlantic expects to generate $58 per month per subscriber for interactive video services: $28 for traditional video, $24 for video on demand and other interactive services (shopping, games), and $6 for other services. Although the $24 for traditional CATV looks reasonable, the $24 for video on demand looks overly optimistic given the high penetration of VCR machines, and video rental stores on every corner.

VIDEO ON DEMAND: WHO NEEDS IT?

Today's home is bombarded with an overwhelming variety of media delivered over an equally overwhelming variety of media. Audio entertainment is physically delivered to our homes on compact discs and audio cassettes and is delivered over the air by FM broadcast radio. We listen to talk shows on AM broadcast radio and obtain the morning news by radio and the physically delivered morning newspaper. Magazines, books, and advertisements are physically delivered each day to our homes.

Video entertainment reaches our homes physically in the form of VCR tapes from the video store and on video discs. Video entertainment reaches our homes over radio waves broadcast over the VHF and UHF spectra and also directly broadcast to us from communication satellites in geosynchronous orbit over the earth's equator. Video entertainment also reaches us over the coaxial cable of the CATV firm. More video entertainment reaches us than we could ever possibly watch, so it is no wonder that so many people feel TV is boring. Yet, according to market data, we continue to watch it for about five hours each day. But is the TV set simply on to keep us company while we do other things, such as read a book or eat dinner?

Video on demand is touted as being more convenient than making a trip to the video store. I suspect, however, that the trip to the video store can be an excuse to get out of the home and that wandering around looking at all the video boxes is actually in itself a form of entertainment and amusement. However, the tape still needs to be returned after its use, and this can be a chore.

My guess is that video on demand will need to cost less than renting a video from the video store. If my conclusion is correct, the Baby Bells are in for a financial fright. The low usage of video on demand in the Rochester, NY trial seems to confirm my conclusion. The Baby Bells clearly believe that consumer spending for video entertainment can be increased greatly, perhaps even fourfold, because of the convenience and variety of offerings from video on demand over their network.

There is yet another financial problem with video on demand. We know from telephone service that consumers do not like usage sensitive pricing where they pay for the amount of time for each local call. Consumers are accustomed to spending money to purchase and own a product or to rent a video tape. However, they do not like to pay each time they view a video.

If going to the video store was a large problem, video delivery services would abound. If returning the tape were such a great problem, video return boxes would appear on every street corner. I can understand that popular videos might be out of stock, which could be frustrating. However, this is one area where technology could help. A popular video could be downloaded to the store over a network and a tape made immediately while you waited, perhaps browsing through other videos and ordering some popcorn.

The other day I looked at the program listing for CATV in my area. The premium CATV channels offered mostly sports, and the basic CATV channels offered more of the same kinds of programs that were available from VHF and UHF broadcasts. To be truthful, I must admit that I do not subscribe to CATV and am more than satisfied by the "free" TV that I receive over my attic antenna aimed at New York City. CATV offers little extra to me, other than more of the same and more of the old same in the form of reruns of old network programs. If over-the-air TV was not available in my area, or if the signals were very weak and of poor picture quality, I would subscribe to cable TV. Many people in New york City subscribe to cable TV for exactly these reasons.

A friend of mine in rural New Jersey is not passed by the cable of cable TV, and hence he subscribes to the 150 channels of direct broadcast satellite TV provided on DIRECTV™. I was overwhelmed by all the channels of movies and other boring programs that were available. The most exciting program I found was world weather. If such program content is what CATV and DBS offer, then there indeed may even be a market for my boring lectures on the life of the sine wave.

A problem with video on demand is that you must know exactly what you want to watch. An advantage of conventional television is that the TV broadcasters think they know what you want to watch.

The disc jockey exists to choose what to listen to for us. You could have access to all the world's music, but you would then have to know exactly what you wanted to listen to. The broadcaster decides the specific music broadcast over radio so that the listener does not have to make any choices. Indeed, there are times when we know exactly what we want to listen to, but many times we are quite willing to allow someone else to choose for us. Of course, the system could program material based on your past listening habits, but for me much of the enjoyment in listening to the radio and CDs is that my patterns change and I like surprises.

One way to approximate true video on demand is to transmit 500 or more TV programs to the home, the concept known as near video on demand mentioned earlier. The problem is how to search through 500 channels to find the one to watch. If you looked at each channel for a half second, then you would need over four minutes would be needed to surf through all 500 channels. Clearly some form of electronic program guide would be needed, but I wonder if looking at the program guide might not be more fun then watching the programs themselves.

One way to use 500-channel TV would be to send the same program starting at different times, so you could watch a program at your convenience. However, most people already use their video cassette machines to record TV programs for viewing at a more convenient time. It is difficult from a consumer perspective to determine the advantage of 500-channel TV, but if it were available consumers might invent a use.

INFORMATION OVERLOAD

All the world's information at your fingertips is what videotex promises, as does also the Internet and the World Wide Web. Access to all sorts of data bases containing all sorts of information, everything you ever want to know, is irresistible. Instant access to people all over the world through e-mail is equally irresistible. However, the vision can be a nightmare too.

We certainly need information. But we are bombarded with information all day at work, and when we return to home, we are bombarded by junk mail and television. If anything, there is too much information. We enter a state of information overload.

The way I obtain the information I really need is through other people, mostly by speaking to them over the telephone. Perhaps I am just old fashioned and a technological Luddite, but my telephone really is my major means of electronic communication. Although my office has facsimile, I do not have a fax machine at my home offices. Nor do I have an address on the Internet. I do not wish to be overloaded with information nor will I allow others to program my time and priorities.

E-mail is a good way to stay in contact with people, particularly for a team of people working together on a common project. However, although e-mail offers you contact with many people all over the world, it also offers all of them contact with you. The problem can again become information overload as you attempt to manage all the unwanted communication. And if speaking to everyone on the phone and reading all your e-mail is too much, imagine looking at everyone who calls you, which is what the videophone would offer.

THE VIDEOPHONE: STILL UNWANTED

I know I am getting old when half my audience has not seen the Stanley Kubrick movie version of Arthur C. Clarke's *2001: A Space Odyssey*. Those who have seen it will remember the picturephone call from the orbiting space station to earth. While at Bell Labs, I was asked to suggest a design for the picturephone equipment used in this movie. So a long time ago, I was a true believer in picturephones as the logical extension of the telephone. Although the design I suggested for *2001* was clearly fiction, AT&T spent $500 million in designing and manufacturing its picturephone of the early 1970s. Why then don't we all have picturephones?

Whenever I speak to an audience about videophones I will ask how many people would like to see the other person while talking with them on the phone. There will usually be a showing of a few hands. I then ask how many people would be wiling to be seen while speaking on the phone. There are hardly any hands raised. This then is the basic problem with the videophone: very few people are willing to be seen while speaking on the phone.

There is an intimacy to speaking on the phone that somehow is invaded by the visual dimension of the videophone. Most people when they want to really pay attention to what you are saying will close their eyes to block out the visual dimension. Thus, videophones simply do not add much to, and most likely actually detract from, a regular telephone call.

The videophone seems to be mostly a product of science fiction, and its only market is science fiction movies and TV shows. Most consumers do not want a videophone—at any price. The negativism is so strong that the only way to make money with videophones would be to offer them as standard telephone service and then charge extra *not* to have them!

Some corporate executives want very much to be seen by their employees and hence use video as a means to broadcast their image to their employees. Such corporate video is typically one-way since the executives have little interest in seeing the employees. I suspect that video teleconferencing is also used by some people who want to project themselves over TV, perhaps in fascination with playing at being a TV news anchor.

TELE-EVERYTHING

Humans are social creatures seeking contact with other humans. Although I usually work at home a few days each week, I find that I need social contact with my colleagues. Hence, I commute to New York City when I am on the East Coast, or to my office at the University of Southern California when I am on the West Coast, to meet with my colleagues. And along the way, I meet other people in various chance encounters.

The technological Utopian vision of people always working at home, telecommuting to work, and maintaining contact electronically solely over the superhighway is greatly exaggerated in its scope. Computer "nerds" might want to avoid social contact, but most normal people crave contact with other people. Technology should never deny humanity but should serve the needs of humanity.

Tele-eduction, distance learning, and the video classroom are likewise over promoted in importance and impact. Students come together to study and learn, and it is this physical coming together that creates a place of learning.

One can imagine that when the book was first invented there were those who predicted the end of educational institutions, since students would now be able to study and learn from a book by themselves without the need to travel to the university. The university of the future will be the university that does not fall prey to an overemphasis on educational technology, but instead uses technology to enrich the educational experience at the university.

Books are disappearing from libraries, not because of technology but because of theft and incorrect shelving. The electronic library ignores the fact that computerized books are difficult to read from a CRT screen and are even more difficult to browse. The problem in many libraries is that the book you want to find is not there. Technology has had great impact in computerizing the card catalogs and now needs to be used to protect books from theft and loss. Computerized searches have increased access to information, but have also made scholars lazy. Furthermore, much material has not been computerized and is not found in the electronic searches.

THE KILLER APPLICATION

When I worked on the AT&T videotex project with Knight Ridder Newspapers, we attempted to present as many different applications as possible to the participants. We did this because we believed that no one single application would be sufficient to justify the service to most users.

The philosophy behind the superhighway takes a different twist and states that a single winning application may indeed justify the whole system. The person who finds that application will make a financial killing—hence the term "killer" application. If there is no such killer application, then it is quite possible that many promoters will instead themselves be killed in their search for the killer application. "The search for the killer application" sounds to me like a title for a horror movie.

The truth is that the killer applications are known but are not acceptable socially. Gambling and pornography are the possible killer applications, but they both have very negative social connotations.

CONSUMER INNOVATION

Consumers sometimes are often better at innovating than the inventors of new technology. One example is the video cassette recorder or the VCR. The VCR was originally conceived as a means for consumers to watch pre-recorded tapes at home and to amass their own video libraries by recording shows off the air. Consumers did all these things but discovered that they could record shows and then watch the show at some other time—the so-called time displacement use of the VCR.

CONSUMER ELECTRONICS

About a dozen years ago, I performed a study of consumer electronics to determine the fundamental driving forces that, in my judgement, seemed to produce positive benefits to consumers and thus account for market success. My study focused on entertainment-based products and services. I examined the history of a number of products and services and their technological evolution. One example is the evolution of the Edison sound cylinder to the 78-rpm disk to the long-playing record to stereo and to today's digital compact disc. Improved quality, improved sensory appeal, long life, and compactness are some of the consumer benefits that emerged from this technological progression.

The consumer benefits that appeared common to most entertainment-oriented consumer-electronics products and services were: sensory appeal, quality, ease of use, portability, personal, program variety, and on-demand use. My guess is that these are still the most important factors to most consumers. Of course, price is also very important, but as discussed earlier, consumers have come to expect lower price and higher quality in consumer electronics as the norm. Based on my list of driving forces, I predicted success for the then fledgling compact disc.

Another aspect of my study was the bandwidth of entertainment-based media compared to communication-based media, such as the telephone. I concluded that entertainment was mostly a broadband process, but that communication for exchanging information and views was a narrowband process. It also seemed that increasing the bandwidth of entertainment media was related to increasing sensory appeal and hence was desirable. For example, in audio, the progression from the Edison cylinder to today's CD has been a progression of ever increasing bandwidth and improved sensory appeal. Telephonic communication is a relatively narrowband process. Attempting to make it broadband through the addition of the visual dimension of the videophone would clearly violate my model and thus failure would be predicted. It is interesting to note that such new communication services as e-mail are even more narrowband than the telephone.

DIFFUSION OF INNOVATIONS

The cumulative market penetration of an innovation follows the so-called "S" curve, because of its shape, shown in the following figure. During the initial market entry, a small number of innovators try the new product or service, and they are then followed by the early adopters. The vast bulk of the market penetration occurs when the early and late majority of consumers adopt the innovation. Finally, the laggards adopt the innovation and the diffusion process is terminated.

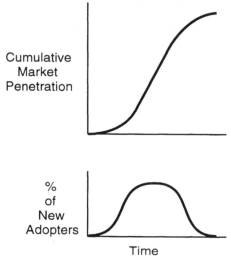

Prof. Everett M. Rogers has studied the diffusion of innovations intensively and has concluded that a number of attributes which emerge in answer to the following questions affect the diffusion process. What is the relative advantage of the innovation compared with older products, services, or methods? Is the innovation truly better? Is the innovation compatible with the values, beliefs, and needs of the adopters? How complex is the innovation to use, understand, and purchase? What is the social status of the innovation? Can it be used easily on a trial basis? Is use of the innovation observable to others?

DETERMINING CONSUMER REACTIONS

How does one determine the reactions of consumers to ideas for a new product or service? The simple way is to ask them. And this leads to a range of market research methodologies for performing the task of "asking," such as questionnaires, interviews, observations, focus groups, and trials. Each methodology has its advantages and disadvantages.

A serious problem with determining consumer reactions to new products and services—particularly those involving the superhighway—is that most consumers have not had any personal experience with the new product or service. Under such unfamiliar circumstances, most people will respond positively, especially if the product or service is described only in positive terms and is positioned as high technology and futuristic. Few people want to appear negative toward technology and the future. There are solutions to these problems, though. The new product or service can be described in a neutral fashion, presenting both positive and negative aspects.

There are a variety of methodologies for determining and measuring consumer reactions. The two broad means of determining consumer reactions are primary research and secondary research. Primary research is the actual conduct of studies and production of data. Primary research is costly and time consuming, but the data is specific to the product or service being evaluated. Secondary research is the examination of data collected by others. Secondary research is inexpensive and fast, but the data is usually not exactly specific to the product or service under evaluation.

Consumer research can be qualitative, in which general reactions are obtained, or quantitative, in which numeric values are assigned to the consumer reactions. The reliability of a study pertains to whether the results can be replicated and make statistical good sense. The external validity of a study pertains to whether the results can be generalized to a large number of consumers. Internal validity pertains to the internal design of the study and whether the results are meaningful measures.

Consumer research can be conducted in a laboratory setting in which all variables are carefully controlled, but a laboratory environment usually is not very realistic. Field research is conducted in the real world and is realistic, but many factors that could affect the results are usually not controllable. Field research is frequently evaluated through observation of the behavior of the participants. Both laboratory research and field research can become contaminated through various factors, such as experimenter bias and improper selection of the participants.

All research designs should be pretested to be certain that the design works and is measuring what is desired. In a manipulative design, the independent variable refers to the factor being manipulated and the dependent variable is the outcome being measured.

Surveys are perhaps the most frequently conducted method of

consumer research. Surveys are conducted through the use of questionnaires. Questionnaires can be self administered, usually on paper by mail or at a site, or they can be administered by a questioner, either in person or by telephone. Questionnaires are composed of a number of questions with responses that can be organized in a check list, ranked using various kinds of scales, compared by preference, or open-ended.

The problem with questionnaires is whether the sample of consumers chosen for the study truly represents the larger external population to which the results will be extended. Bias can result if some people are more willing to respond than others. The size and constitution of the sample needs careful consideration. The questions themselves need very careful design and adequate pretesting to be certain that they are understood and produce responses that are accurate measures of the true feelings of the respondents. How a question is asked can bias the results. For example, asking people whether they want to see the other person during a telephone call can generate very different responses from asking people whether they are willing to be seen during a telephone call.

A technique frequently used to obtain consumer responses to a new product or service is to use a focus group. A small group of consumers, usually chosen to have similar demographics, are assembled in a room and are lead in a discussion about the new product or service by a professional moderator. The sponsors of the study typically are in a separate room separated from the interview room by a one-way mirror so that they can not be seen by the participants. Each participant in the focus group can hear the comments of others and can react themselves.

While the results of focus groups are not statistically significant, a focus group can give an overall impression and suggest ways to market and position a new product or service. Focus groups are subject to the whims of the moderator, a single vocal participant can sway the whole group, and the small number of participants might not represent any larger population.

Another means to determine consumer reaction to a novel product or service is to offer the product or service in a trial. Real consumers then will have real experience in actually using the product or service. The consumers might actually be asked to pay real money for the use, thereby giving real market information. However, all the promotion and attention given to most trials creates a sense of positivism in most consumers, thereby skewing the results positively. It takes careful design and attention to avoid the problems of trials, along with independent evaluation. So important are the pitfalls of trials that a later chapter is devoted solely to them. Many trials achieve a life of their own, in which the people implementing the trial would be the last to terminate their own jobs by presenting a negative finding.

Most of today's superhighway trials are technology trials testing whether the technology will actually work or not. The market trials conducted thus far use small populations of consumers thereby raising questions about how the participants were chosen and how representative they are of the population of normal consumers.

BUSINESS AND THE CONSUMER

We all probably agree that the consumer is the most important factor in shaping and determining the future. However, any new product or service is developed and provided to consumers by businesses. The culture of the business entity, corporate strategy, the pricing and billing mechanisms, the commitment to service, and other such aspects of the business comprise the last factor in shaping the future.

READINGS

Asthana, Praveen, "Jumping the technology S-curve," *IEEE Spectrum*, June 1995, pp. 49-54.

Carey, John, "The Market for New Residential Services," *Broadband Networks*, ed. Elton, Martin C. J., North-Holland: New York, 1991, pp. 9-22.

Rogers, Everett M., *Diffusion of Innovations (Third Edition)*, The Free Press: New York, 1983.

Williams, Frederick, Ronald E. Rice, & Everett M. Rogers, *Research Methods and the New Media*, The Free Press: New York, 1988.

Chapter 8

BUSINESS CONSIDERATIONS

The most wonderful new product or service is of little value without an appropriate business to provide it to the public. In this chapter, we examine some of the various business considerations that are necessary for success and that help shape the future.

BUSINESS STRATEGY

"Strategy" and "strategic" are much in vogue today, and even universities are attempting to think and plan strategically. But an objective comes before a strategy. A strategy is simply the plan and actions to achieve the objective. And objectives should be consistent with the mission of the organization. So we have mission, objective, and strategy. It certainly sounds simple and should be, but I have seen many lengthy and meaningless mission statements.

A few years ago, an op-ed piece that I wrote about the AT&T acquisition of NCR was published in *The New York Times*. In it, I accused AT&T of having a confused corporate mission and suggested that AT&T should divest manufacturing and concentrate on network services. In its response, also published in *The New York Times*, AT&T's chief strategist, Richard Bodman, disagreed with my advice and criticism, and stated that AT&T's "mission is broad and deep: to be the world leader in information technology." My colleagues and I were aghast. The mission statement was even more confused than we had thought. How do you measure "world leader"? What is "information technology"? Both these terms are so "broad and deep" as to be meaningless and unmeasurable. Later in this chapter, we shall see that AT&T ultimately decided to divest manufacturing in 1995.

Many companies and universities do not really understand strategy, mission, and objectives. The mission statements are apple-pie and motherhood with unmeasurable objectives and confused lengthy strategies. AT&T has become more focused today and has concentrated on providing communication service. The announcement in 1995 of its planned divestiture of its manufacturing businesses is consistent with this more focused mission as also was the acquisition of McCaw's cellular business. However, the McCaw acquisition places AT&T in direct competition with the cellular ventures of the Baby Bells, and perhaps this was a factor in the clamoring by the Baby Bells to be allowed to provide long-distance service in competition with AT&T.

Before the mergers of some of the Baby Bells, GTE was the largest regional holding company providing local telephone service in the United States. GTE used to be a mini version of the Bell System with its own manufacturing arm, Automatic Electric, and its own long-distance company, Sprint. But GTE divested its manufacturing and long-distance operations, and also its Sylvania lighting division. GTE developed a clear strategy to focus on the provision of local telephone service. Given the large profit margins of local service and also GTE's excellent profitability, this strategy has clearly paid well for GTE. However, there are murmurings that GTE is thinking of re-entering long distance.

Pacific Telesis decided to concentrate its efforts on providing local telephone service, and thus divested its cellular and other ventures into a new company, AirTouch Communications. In 1995, Pacific Telesis had a profit margin of 12 percent, but AirTouch's profit margin was only 8 percent. Telesis's cash flow margin was 43 percent versus AirTouch's 20 percent. Clearly, the provision of local telephone service is a very profitable enterprise and focusing on it makes good sense.

US WEST has separated itself financially into two businesses: one for its local telephone business and the other for its ventures into entertainment and multimedia. This might ultimately lead to a separation of local telephone service from the other ventures, which thus far have been much less profitable. Of course, the hope is that the cellular and new ventures, though risky, will ultimately become great financial successes. But for 1995, the US WEST Media Group—the new ventures company—had a profit margin of only 6 percent versus the profit margin of 12 percent for the US WEST Communications Group—the telephone company. US WEST is not alone in discovering that the provision of local telephone service is very profitable—much more so than entertainment and multimedia.

THE BABY BELLS' PURSUIT OF STRATEGY

All the Baby Bells became linked with each other in various combinations through a series of joint ventures in the cellular and entertainment businesses announced in 1994 . Bell Atlantic, NYNEX, and Pacific Telesis joined together in 1994 to develop home entertainment, information, and interactive services with assistance from Creative Artists Agency, but the venture was dissolved a year later. Ameritech, Bell South, and SBC Corporation (formerly Southwestern Bell) joined in 1994 with the Walt Disney Company to develop new video entertainment services. Pacific Telesis divested its cellular and other new ventures in 1994 to create AirTouch which then linked that year with US WEST in a joint cellular venture. Bell Atlantic and NYNEX also joined in pursuit of a linked cellular venture, while a higher level linking of the Bell Atlantic/NYNEX and AirTouch/US WEST cellular ventures was also being contemplated.

At the end of 1995, Bell Atlantic and NYNEX were reported to be considering a merger, perhaps also with Pacific Telesis, to provide seamless local and long-distance telephone service. In February, 1996, SBC announced its plans to acquire Pacific Telesis. In April 1996, Bell Atlantic and NYNEX announced their intent to merge into a single company.

One needs a map to follow all these joint linkages, mergers, and acquisitions. All this linking simply makes the relationships of the Baby Bells very cozy with other and decreases any real chances for competition between them.

US WEST has been the most active of the Baby Bells in pursuing the entertainment business. In 1992, US WEST joined the CATV-giant TCI and AT&T in a trial of viewer-controlled CATV. In 1993, US WEST purchased a 25 percent interest in Time Warner Entertainment. In 1995, US WEST created two US WEST companies, each with their own stock, one for telephone service and the other for entertainment and other ventures. Not only US WEST, but many of the Baby Bells are pursuing various new ventures involving the provision of entertainment and of CATV service. We must ask what they really know about the entertainment business, although their antics have certainly been entertaining.

The Baby Bells one day clamor to be in manufacturing, the next day in long distance, and then in video entertainment. What they do not seem to want to provide anymore is local telephone service. When Theodore Vail and his AT&T gobbled up the local phone companies at the beginning of the century, AT&T provided central direction and apparently kept the Bell companies under control. Now that the Baby Bells are free of AT&T, they seem to be struggling for a sense of purpose and direction. The glamour and glitz of Hollywood has indeed stricken them with media mania.

But it is not only the Baby Bells who seem to have Hollywood fever. Many other firms are lusting after Hollywood, convinced that great pots of gold are at the end of the media rainbow. We therefore are seeing many mergers and acquisitions as many grab a partner before it is too late. Nobody wants to take a chance of being left outside in the cold when the warmth of sunny southern California and Hollywood beckons. We are reminded of Sony's and Matsushita's excursions through sunny California and of the billions of dollars they both lost on their purchases of studios. Yet the frenzy increases as computer software giants and newspapers join the party forming their own strategic alliances, mergers, and acquisitions.

Although mergers and acquisitions seem to be gripping the communication industry, divestitures are also still popular. We already mentioned Pacific Telesis' divestiture of its cellular business to create AirTouch. In the next section we will examine AT&T's "breakup II."

Given all the nonsense, fear, and panic that seem to be gripping the communication industry, the best strategy is to sit on the sidelines and wait. Nothing will happen quickly, other than big losses. When the future becomes more certain is the time to act and join with the few successes.

AT&T BREAKUP II

In 1990, I performed a financial analysis of AT&T and concluded that AT&T's manufacturing business was losing money and dragging down AT&T's overall profitability. My calculations showed AT&T was losing from $1.5 billion to as much as $5 billion from the sale of products, although these estimates did not include any internal sales within AT&T. I concluded that AT&T should divest its manufacturing business and concentrate on providing telecommunication services such as its profitable long-distance service.

When AT&T announced its intentions to purchase NCR in early 1991, I was baffled that AT&T would want to increase further its exposure to the losses and low margins of manufacturing. As mentioned earlier, I wrote an op-ed piece in *The New York Times* stating that "AT&T should probably be unloading its losing product business" and that "a proper strategy for AT&T would involve developing and articulating a corporate vision to make the company the leading force in telecommunication services." AT&T did not then follow my advice. AT&T continued to believe that there was some form of synergism between telecommunication services and computing. In a response to my op-ed by AT&T's chief strategist, Richard Bodman, AT&T claimed that "products and services are increasingly hard to separate."

Clearly within four years AT&T changed its mind and finally followed my advice with its announcement in the Fall of 1995 that manufacturing and NCR would both be divested by 1997. It does take courage finally to take action that admits and corrects a mistake. I guess it is "better late than never," although I hate to think of the financial losses that could have been avoided over the four years.

AT&T will break apart into three major companies: a new AT&T dedicated mostly to providing telecommunication services, such as long distance and cellular; a telecommunication products company, to be called Lucent Technologies; and a computer company, to be called NCR. AT&T further announced that it will sell its majority ownership of the AT&T Capital Corporation and use the proceeds to reduce debt.

My financial analysis of AT&T Breakup II is based on figures from AT&T's 1994 Annual Report and also from some details of the announced breakup. The Annual Report gives revenues for many of AT&T's separate business along with direct costs. However, research and development (R&D) and selling, general, and administrative (SG&A) expenditures are stated only from a corporate perspective and are not allocated to the separate businesses. The trick is how to perform these allocations so that each business can be assigned all its costs and then reassembled according to the new breakup.

R&D expenditures for 1994 were $3.1 billion, which includes AT&T Bell Labs and NCR R&D. AT&T Bell Labs states that 4,000 of its 26,000 people will be assigned to the new AT&T, and that the remaining 22,000 will be assigned to Lucent Technologies. If we assume that NCR's R&D

is a high 11% of revenue, then R&D for NCR would be about $0.5 billion, leaving $2.6 billion to be allocated across the new AT&T and Lucent Technologies according to their share of the 26,000 Bell Labs people. This works out to $0.4 billion for the new AT&T and $2.2 billion for Lucent Technologies.

AT&T's SG&A expenses for 1994 was a large $19.6 billion. The allocation of SG&A expenses across AT&T's various businesses clearly has a large impact on the net profitability of each business. I investigated two approaches to performing this allocation. The first method allocated SG&A expense in direct proportion to the revenues of each business after eliminating access charges. Access charges are collected as revenue by AT&T but are then paid as an expense to the local telephone companies. Access charges in 1994 were a high 41% of AT&T's telecommunication services revenue and would overly skew the allocation of SG&A if they were included in revenues.

The second method allocated SG&A in direct proportion to the direct costs of each business, again net of access charges. This approach is perhaps more meaningful than allocating by revenue since a business with high direct costs will probably also have high SG&A expenses.

A major surprise from the analyses is that no matter how SG&A expenses are allocated, NCR is profitable, with a 4% estimated profit margin using revenue allocation and a 6% profit margin using cost allocation. At the time of its acquisition by AT&T in 1991, NCR was just barely profitable with a downward trend. It indeed is a miracle if AT&T somehow had made NCR more profitable.

I had concluded in my 1990 analysis that Western Electric was a financial drain on AT&T. The new analyses of AT&T breakup II confirm this conclusion. Depending on whether SG&A is allocated by revenues or by costs, Lucent Technologies' estimated after-tax profit margin is only a meager 1%, or a large loss of 10% on revenues of $20 billion. This is not a good picture either way. The corporate symbol for Lucent Technologies is a red, blurred zero—which is very appropriate given my prediction of large losses.

The new AT&T will be a very profitable company, with an estimated profit margin of either 8% or nearly 14%, depending on how SG&A is allocated. As a base of comparison, the more profitable Baby Bells had an after-tax profit margin of 15% for 1995. The new AT&T clearly will be more profitable than its major competitor, MCI. Will this then lead to increased price competition in the provision of long-distance service, since the new AT&T will no longer be held back by its less profitable manufacturing businesses?

The acquisition of McCaw's cellular business by AT&T makes much strategic sense and strengthens AT&T's service business. However, McCaw has not yet contributed to AT&T's profitability. Although McCaw created $2,062 million in revenue for the last nine months of 1994 when it was owned by AT&T, McCaw contributed only $34 million in profits. In fact, AT&T's profit margin would be higher if it abandoned the cellular business, although such a move might be a serious strategic error.

At the time of the 1984 divestiture of the seven Baby Bells from AT&T, I wondered whether it would have made more sense for AT&T to divest manufacturing and keep the end-to-end provision of telephone service intact. I still wonder whether it was a mistake for AT&T to have divested itself of the Baby Bells, particularly since they now want to compete with AT&T in the provision of long-distance service. Manufacturing will now finally be divested from AT&T, but not in reaction to any government pressures, although the government has been attempting to pry manufacturing from AT&T for nearly a century. One wonders why AT&T held onto manufacturing for such a long time. This divestiture is indeed ultimate since there will now be little left for AT&T to divest further.

In its public statements about the new breakup, AT&T has attempted a positive spin and does not admit to any financial losses from manufacturing. Indeed, AT&T had become schizophrenic internally since the Baby Bells were both a customer in their equipment purchases and a potential competitor in long distance. It will be interesting to see whether AT&T enters the local market, perhaps in collaboration with CATV companies, to compete with the Baby Bells on their home turf. Perhaps the easiest way for AT&T to enter the local market would be to acquire an already existing local telephone company which has national presence. GTE meets that profile.

The Baby Bells have been lobbying over the years to be allowed into manufacturing. Given the low profitability—and perhaps large losses—of Western Electric over the last decade or more, one wonders how wise the Baby Bells are in this strategy. If they still want so strongly to enter manufacturing, perhaps they will make an attempt to purchase Lucent Technologies once it is divested from AT&T.

A century ago, Theodore N. Vail, the founder of the Bell System, saw the wisdom of long-distance service as the glue to keep together the Bell System. Vail's vision glows even more strongly today as AT&T returns to being mainly a long-distance service provider.

R&D COMMITMENT

The telecommunication industry has been built progressively and steadily on major accomplishments in the progress of science and technology. The former Bell Labs was supported generously by AT&T and created the knowledge base for much of this progress. To its credit, AT&T continued its substantial commitment to basic research at AT&T Bell Labs during the twelve years after the Bell breakup. Much of the results of this research becomes available to the entire telecommunication industry, thereby benefitting the United States as a whole.

With the Bell breakup in 1984, a new R&D facility, Bellcore, was created to serve the Baby Bells with all seven as equal owners. Since then, most of the Baby Bells have created their own internal science and technology centers, although I personally have seen little of value from these small and unfocused efforts. After years of rumors, it was finally announced that the Baby Bells will sell Bellcore. This tells me that the Baby Bells have little or no commitment to basic research in communication.

I see now how wise it was for AT&T to collect a small percentage of the revenues of the local telephone companies and then use these funds to support basic research. Without AT&T's wisdom and commitment to longer term priorities, the Baby Bells seem doomed to a pursuit of short-term profits and to media mania, and they do not care about research.

The further breakup of AT&T that will occur in 1997 will ultimately destroy basic research at Bell Labs since the new manufacturing business will not be able to justify this costly expenditure, given short-term profitability concerns. This elimination of industrial support for basic research in telecommunication is indeed is a sad day for the United States as a whole.

CONVERGENCE AND SYNERGIES

Decades ago the term "convergence" entered the business lexicon. Computers and telecommunication were converging according to the believers and proponents of convergence. The use of computers to control telecommunication switching systems and the use of telecommunication to gain access to computerized data bases were the two trends behind these claims of convergence. Business consultants then stated that to be adequately positioned for the future, a firm could no longer rely on being solely a computer company or a telecommunication company. Accordingly, IBM acquired Satellite Business Systems, and made a big mistake. AT&T attempted to develop its own personal computers, and made a big mistake. AT&T attempted to salvage something from its continuing belief in being in the computer business through an ill-fated involvement with Olivetti and then finally through the acquisition of NCR. NCR will be divested from AT&T in 1997 thereby finally ending AT&T's involvement in the computer business.

A few years ago, business consultants concluded that a hardware company had to be in software also because of supposed synergies between the two. Based on the views of the synergists, the Sony Corporation acquired Columbia Pictures and the Matsushita Electric Industrial Company acquired MCA. The synergies never appeared, and both Sony and Matsushita lost billions of dollars in the deals.

Manufacturing consumer electronics is a very different business than making movies. Manufacturing personal computers is a very different business than providing telephone service. The finances are quite different, but more importantly, the management styles and basic business skills are very different.

MANAGEMENT STYLE

Different businesses usually are associated with different styles of management. The management of telephone companies has traditionally been quite conservative. Management of a monopoly telephone company does not entail much risk, since there really is not much competition. It does however involve much lobbying with regulators and politicians. Working for the telephone company used to be a job for life, but advances in technology have resulted in dramatic increases in productivity, thereby resulting in much downsizing for many telephone companies.

The entertainment industry has a management style that is nearly opposite that of a telephone company. There is considerable acceptance of risk on the part of management in the entertainment industry. Most movies never recoup their costs, although the trick here is to risk investor money and never your own. There is much glamour and glitz in the entertainment industry, particularly on the production side. The public has a fascination with the stars of the entertainment industry, and perhaps the investors who lose their money obtain some gratification from rubbing elbows with the stars. The CATV industry has a management style somewhere in-between the telephone industry and the production portion of the entertainment industry.

Clearly, the entertainment business is quite different from the telephone business. Hence, if telephone companies were to enter the entertainment business, the creation of a new subsidiary to manage the entertainment ventures would seem to make good sense, and would also facilitate clear analysis of financial performance. Any possibilities of cross subsidization would also be eliminated. Investors with different objectives would also know more clearly what kind of business they were investing in.

Another aspect of management style is a willingness to acknowledge and accept differing views. The Baby Bells appear to have fallen for their own hype about the superhighway. I wonder who is advising their top management and how much diversity in opinion is accepted within the companies. GTE California's overly protective reaction to my critical comments of its Cerritos trial would imply to me that diversity in views and critical thinking are not much in vogue at GTE.

Years ago, AT&T's corporate planning department was characterized by much independent thinking and by differing views, but much of this independence unfortunately vanished before the breakup of 1984. Much of my time at the AT&T Marketing Department was as an in-house loyal critic, and my views were valued by my immediate management. Some companies can realize the importance of independent thinking within the organization. Decisions are best made as possible when as many differing views are openly discussed and considered.

Today's telecommunication industry uses very sophisticated technology, and I doubt that the top executives of many companies themselves have the engineering and technological background and knowledge necessary to understand the technology and its implications. This means that they must rely upon others within the corporation. Unfortunately, many managers have become proponents and promoters and are more interested in their own short-term advancement than the longer-term future of the corporation. The result is self-serving advice that is often not the correct course of action. Simply explained, there is much confusion, panic, and fear at the top levels of management of many of the players in the drama of the superhighway.

PROCESS MANAGEMENT

In the late 1970s, the AT&T Marketing Department invented a process for evaluating and bringing new products and services to market. The process was called the Phased Management Decision System, or PMDS for short. It consisted of seven steps with detailed requirements for documentation along with numerous phase review committees to evaluate each step. A number of thick binders described in detail the various steps and required supporting documents. The flow chart of the process was huge and covered many pages. It was clear to me that nothing would ever make it through such a complicated and overly detailed process. Such are the pitfalls of process management. It took a few years before it was realized that not a single new product or service ever made it through the complexities of the PMDS process, and ultimately PMDS was scrapped for a far simpler process.

Indeed, a process for evaluating and bringing new products and services to market is needed. However, the process must be simple and easily understandable. Supporting documentation is also needed, but it too should be short and understandable. The problem with AT&T's PMDS, and similar such overly elaborate systems, is that all the emphasis is on the process and not on the new product or service. No longer is the product or service being managed, but the process becomes an end in itself. Committees and elaborate documents become the end objective, and the objective of evaluating and bringing a product or service to market in a timely fashion is forgotten. I recall remarking in my criticism that PMDS had become the end product and that no matter how nicely we packaged PMDS, customers would not buy it.

While at AT&T, I produced my own version of the business process. The process, shown in the two diagrams at the end of this chapter, begin with a synergistic matching of technology and consumer needs from which ideas are developed for new products and services. The marketability of these ideas is then tested and evaluated, and the results are fed back to impact the technology and consumer factors. The ideas that are marketable create a vision of the future which then results in a strategy and plan to achieve that vision. Business cases, evaluations, and decisions need to be made along the way in an iterative fashion as more information and results are obtained. The final development process brings the product or service to the marketplace. The results of performance in the marketplace then modify the plan in an iterative fashion.

An overemphasis on process management usually occurs when management is insecure and does not fully understand technology, products, services, or markets. By emphasizing the process, management feels secure that matters are under control. But managing the process then becomes the major objective—not bringing products and services to market in a timely fashion. The process and its management can also become a weapon used to delay and destroy ideas and their advocates. In the end, though, such unwieldy processes usually die of their own weight, but along the way many good ideas and people can be destroyed.

IMAGE

The image of a company is an important factor in helping to determine what types of new businesses it might enter successfully. Telephone companies have had an image of being non-responsive to customers and of not really caring about customers. The usual response to a repair or service call to the telephone company is that they will do it their way when they decide. Who has not waited at home for hours for the telephone repair or installation people? On weekends and evenings, most telephone company business offices are closed. I'm reminded of the old adage: "We're the phone company, we do it our way." However, telephone companies do seem to have a positive image when it comes to the quality and reliability of telephone service.

MANAGING THE CUSTOMER INTERFACE

While conducting research for this book, I telephoned many different firms. Along the way, I learned a lot about how different firms manage the customer interface, and in particular, how **not** to manage this interface.

My phone calls to many companies were answered by computers

that then presented all sorts of options accessible by pushing the appropriate button on my touchtone phone. I found the long lists of options to be very confusing, and in some cases was never presented with an option that fit my particular case. When I finally did reach a human, some firms passed me from person to person as if my request for information were a hot potato. In calling two long distance companies to obtain information on their packet data services, I never did reach the right person and ultimately gave up.

I recently responded to an advertisement in a major magazine for an investment company by dialing the toll-free number that was given. I obtained a recording that stated I had reached a nonworking number at the company. I finally reached an operator and asked to speak to the president of the company, but was asked which president, the New Jersey president or the New York president. Clearly, I would not trust my money in a company riddled with such chaos and confusion.

The people who answer the phone at a large company represent that company to the public and should be adequately trained and know the company's basic products and services. In the case of my attempts to reach someone familiar with packet switched data, many of the people I talked to at two large long-distance companies had no idea what I was even talking about. Recently, I was unable to obtain accurate information about some of the features of AT&T's voice-message service and finally, after much finger-pointing between AT&T and Bell Atlantic, had to discontinue the service in order to make my phone work correctly.

Although training and education can be costly affairs, they are essential. I remember that when I worked at the AT&T Consumer Products division, we were planning to introduce a modem for personal computers. The problem was that our sales force knew nothing about modems, since their sales experience had been restricted to telephones. I put together a short lecture on the basics of modems, and we videotaped it for showing to the sales people. The skills, knowledge, and competencies of the employees of a company determine what the company can do realistically with new products and services.

A telephone call to many companies today is a frustrating affair. You are immediately intercepted by the computer with its recording asking you to enter a "1" if you have a touchtone phone. If you are like me, you have learned never to do that but instead to wait for a human. Otherwise you are led through ever decreasing depths of a complex search to reach a human. Sometimes you never hear what you are listening for and are tossed back out of the loop to call back again for another attempt.

I have this image that some companies have downsized so much that there are no humans left—just a computer answering the phone and connecting callers to voice mail.

Billing is an important aspect of the customer interface. AT&T has recently introduced direct billing for its larger residential accounts, but still obtains much of its billing through the local telephone companies. This is a clear mistake since AT&T has relinquished its identity to the local telephone company. Both the CATV operator and the local telephone company have monthly billing systems in place. Telephone service is measured but CATV service is not. The telephone bill thus is more complex and lengthy.

SIZE AND COMPLEXITY

A company can become too large, so that employees can no longer comprehend all the various different businesses that comprise the overall company. This is also true for top management. A chief executive can only be competent and knowledgeable in just so many areas. The more complex the company in such aspects as its organizational structure and different businesses, the more management and employee confusion that ensues.

The management style and knowledge required for a manufacturing business is very different from that for managing a Hollywood studio, as Sony discovered with its ill-fated acquisition of Columbia Pictures. The Baby Bells plead to be allowed into manufacturing and entertainment, yet their management knows little about these businesses. Disaster most likely would be the result, but since the Baby Bells provide basic telephone service, which is an essential utility, the Nation can not allow the risk of such a disaster to occur.

CUSTOMER RESPONSIVENESS

Responsiveness is important in the telephone business given its essential nature in time of an emergency. The repair office of the telephone company operates 24 hours each day and is always available. The telephone company business office is open during the business day and on weekends. Some telephone companies maintain evening hours at their business offices for the convenience of their customers who work during the day. Most CATV operators maintain 24-hour access to their repair lines, but their customer service varies widely. Some are available on a 24-hour basis, but other close for minor holidays. AT&T is always available 24-hours each day, demonstrating a true commitment to its customers.

Telephone companies used to be notorious for missing repair and service appointments. The result is that in some states they are required to be on time, and, if late by some specified amount, must compensate their customer for missed work hours and other losses. CATV companies were equally notorious for lateness, and now many CATV operators offer a free month, or $20, if they fail to meet an appointment by more than some specified margin. But there is a cost associated with providing a high level of customer responsiveness. Those who favor competition in the provision of local telephone and cable services will point out that with many competing providers there can also be many different levels of service, and those consumers who want the highest level will have to pay accordingly.

THE POLICY ARENA

Are local telephone service and CATV service natural monopolies? Does competition make sense and can it replace regulation? How will universal service be provided to all in a competitive world where some customers might be far more costly to service than others? What should be the nation's policy in determining answers to such questions? The answers to these questions bring us to the last of the five links, namely, policy and regulatory issues.

READINGS

Bodman, Richard, "The Word on Strategies From A.T.&T.", *The New York Times*, April 14, 1991, Section 3.

Grover, Ronald and Mark Landler, "Hollywood Scuffle," *Business Week*, December 12, 1994, pp. 36-38.

Noll, A. Michael, "The Failures of A.T.&T. Strategies," *The New York Times*, March 31, 1991, Section 3, p. 9.

Rebello, Kathy and Paula Dwyer, "Guess Who's Out Front on the I-Way," *Business Week*, December 19, 1994, pp. 114-116.

Weber, Jonathan and John Lippman, "Interactive TV: A Thorny Horn of Plenty," *Los Angeles Times*, June 14, 1993, pp. D1-D2.

"America's television industry: Incest is best," *The Economist*, April 17, 1993, p. 66.

"Fatal attraction," *The Economist*, March 23, 1996, pp. 73-74.

"Why first may not last," *The Economist*, March 16, 1996, p. 65.

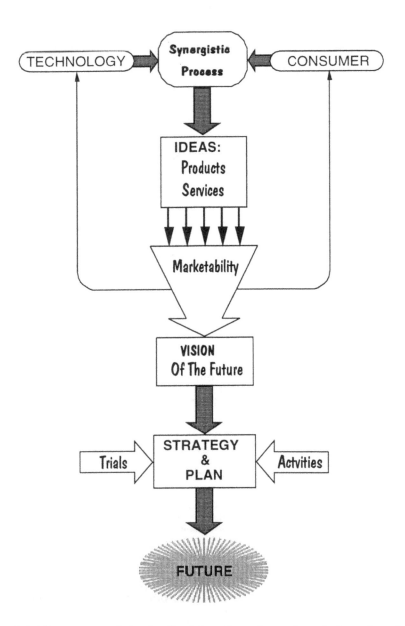

Fig. 8-1. The process of shaping the future through novel products and services begins with a synergistic interaction between technology and consumer needs. The marketability of these ideas then determines a vision of the future which can be achieved through an appropriate plan and strategy.

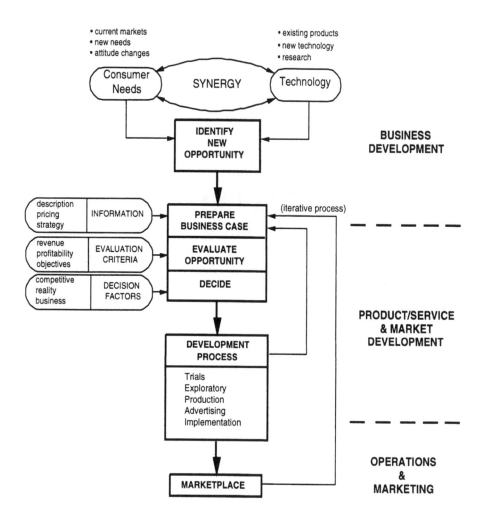

Fig. 8-2. The overall process of bringing a new product or service to market involves much evaluation and iteration. The process shown here consists of three major steps, labeled on the right.

Chapter 9

POLICY AND REGULATORY ISSUES

Telecommunication in the United States has traditionally been subject to regulation at the federal, state, and even municipal levels. This means that what may be possible technologically might nevertheless encounter regulatory and policy barriers.

In this chapter we examine the fifth and last link in determining the viability of a new venture, namely, policy and regulatory issues. Remember that any venture will only be as strong as the weakest of the five links. Although some of the five might seem less interesting than others, they are all equally essential. We start our treatment of policy and regulatory issues with the creation of regulation for telecommunication and end with a discussion of the issues affecting the superhighway and video dial tone.

THE CREATION OF TELECOMMUNICATION REGULATION

In the distant past toward the end of the nineteenth century, after the expiration of the Bell patents on the telephone, a number of local telephone companies formed to compete with the Bell companies. By 1903, the number of non-Bell telephones significantly outnumbered Bell telephones. In 1907, the founder of the Bell System, Theodore N. Vail, returned to the Bell System as president of AT&T after a twenty-year absence. Vail believed that the provision of telephone service was a natural monopoly and that competition made no sense. But he realized that unbridled monopoly would not be tolerated by the public and government.

Vail's solution was to endorse regulation at the state level to protect the public from the abuses of monopoly, although the cynic would claim that Vail only endorsed regulation because he knew that the Bell companies could dominate the regulatory commissions. The state commissions and various municipal regulators, from whom the local telephone companies obtained their franchises to use the public right of way, were the only regulators of telecommunication for nearly three decades. The Communications Act of 1934 created a federal body to regulate telecommunication in the United States, namely, the Federal Communications Commission, or the FCC.

The FCC was created as a form of adversary to AT&T and thus was opposed in principle to the Bell monopoly. For decades, the FCC has been attempting to force the introduction of more competition in the provision of telecommunication.

In 1962, AT&T launched the first telecommunication satellite, *Telstar.* A year later, the first geosynchronous satellite for intercontinental telecommunication, designed by Hughes Laboratories, was launched. The FCC, under lobbying pressure from the aerospace industry, became concerned that AT&T would monopolize this new technology. As a result, a new entity, Comsat, was created to control communication satellites. AT&T was allowed no more than 27.5 percent ownership of Comsat. In 1972, the FCC pressured AT&T to dispose of all its interests in Comsat, thereby forcing AT&T from the communication satellite business.

The history of cellular service is yet another example of the FCC's influence in promoting competition in the provision of telecommunication service. The technology for today's cellular telephone service became available in the early 1970s, but it was not until 1982 that the FCC finally decided how to create competition in the new cellular industry. Hence, the availability of cellular service in the top markets was delayed until 1984. The FCC's solution was that there be two providers: the so-called wireline company (usually the telephone company) and a non-wireline company.

Although given a congressional mandate to regulate telecommunication in the United States, the FCC seems to have developed a philosophical objection to regulation, and thus, seems to prefer not to regulate any more. The FCC's response to many issues, such as the need for standards for AM stereo, is to "allow the marketplace to decide," or more recently, to auction off the radio spectrum for new, but, yet nonexistent, personal communication services (PCS). We will have to see whether this anti-regulatory stance of the FCC is wise or not. The FCC, however, certainly was out to dismember AT&T and foster competition in the provision of long-distance telephone service.

After a lengthy investigation conducted by the FCC and interrupted by World War II, the United States Department of Justice, in 1949, charged AT&T with violations of the Sherman Antitrust Act. The case dragged on and was finally settled on January 12, 1956 with a Consent Decree between AT&T and the Justice Department. AT&T agreed to limit itself solely to the provision of common-carrier communication service, and in return the Bell System was left intact. This Final Judgement would later be modified in 1982 and result in the actual break-up the Bell System.

THE BELL BREAKUP

Perhaps the single greatest impact on the future of telecommunication in the United States was the breakup of the Bell System that was mandated in 1982 and that occurred on January 1, 1984. The forces that led to this breakup were competition and the anti-competitive actions of the Bell System. AT&T agreed to the breakup and helped in its design to be free of the restrictions of the 1956 Consent Decree. The 1956 Final Judgement restricted AT&T to the provision of telecommunication and in effect prevented AT&T from entering the computer business. AT&T, however, had come to believe that it had great opportunities in the computer business and thus wanted to be free of these restrictions.

The details of the breakup, which were agreed to by AT&T and the federal Department of Justice, were stipulated in the Modification of Final Judgement (the so-called MFJ). Judge Harold C. Greene presided over the antitrust case against the Bell System, and his court had to concur with the final deposition of the case as decided by the Department of Justice and AT&T.

The major stipulations of the MFJ were the divestiture of the local Bell operating companies (BOCs) from AT&T and the restriction of the BOCs solely to the provision of local exchange telephone service within their operating jurisdictions. All previous business restrictions on AT&T were lifted. The wording of the MFJ was clear in its intent to punish the BOCs and to prevent them from ever again using their local monopoly power for anti-competitive purposes.

Under the terms of the MFJ, the United States was organized into a number of Local Access and Transport Areas (LATAs), and the Baby Bells were forbidden to carry telecommunication traffic between these LATAs, even if the LATAs were within their state. All inter-LATA traffic was to be carried only by long-distance companies, such as AT&T, MCI, and Sprint. The punishment inflicted upon the BOCs was that intra-state long-distance was taken from them and given to the long-distance carriers.

AT&T had attempted to retain the lucrative directory business, but Judge Greene instead allowed the BOCs to retain this business. There also was a battle over the use of the Bell name, and in the end Judge Greene allowed the BOCs to retain their Bell identity and AT&T was forbidden from use of the Bell name other than for AT&T Bell Labs.

The actual divestiture of the BOCs from AT&T was a very complex affair but was accomplished in less than two years from the final approval of the MFJ in August 1982. The BOCs were organized into seven regional companies which then owned the local BOCs. These seven regional holding companies are today called the Baby Bells, and many of them have reorganized to eliminate the identity of their local operating companies. In New Jersey, for example, New Jersey Bell has now become Bell Atlantic New Jersey to emphasize the identity of the regional holding company, Bell Atlantic. AT&T retained its manufacturing arm (previously, Western Electric) and its long-distance business and was free to enter the computer, or any other business.

The major reasons that AT&T agreed to the breakup were AT&T's strong beliefs that its future was in computers and that the provision of local telephone service was no longer a growth business.

THE BABY BELLS: THE FINANCIAL SUCCESS OF DIVESTITURE

Perhaps the biggest surprise of divestiture was the financial strength of the seven Baby Bells. In 1982, AT&T had a net income of $7.3 billion and an after-tax profit margin of 11.1 percent. In 1984, the first full year of divestiture, AT&T's net income shrank to $1.4 billion with a profit margin of only 2.6 percent. The seven Baby Bells had an aggregate net income of $6.8 billion and an overall profit margin of 11.8 percent for 1984.

Clearly, the local telephone companies were the financial strength of the old Bell System. AT&T's manufacturing arm, Western Electric, might have been profitable in the distant past, but it was no longer profitable and instead had become a financial liability to AT&T. AT&T's attempts to create its own computer business were dismal failures, and ultimately, in an attempt to salvage something, AT&T acquired NCR. But even with this acquisition, AT&T's computer business continued to be a drag on AT&T's overall profitability. Meanwhile, as we saw in an earlier chapter, five of the Baby Bells in 1995 had profit margins of 11 to 15 percent. The least profitable Baby Bell, NYNEX, had a profit margin of 8 percent for 1995, which exceeded AT&T's profit margin for 1994.

It took a few years for the ashes of divestiture to settle and for the Baby Bells to realize their financial strength. The Baby Bells created a few new business ventures in the areas allowed under the restrictions of the MFJ. Some of these ventures were outright failures, such as NYNEX's and Bell Atlantic's forays into retail computer stores. A number of the Baby Bells made international investments in cellular service, telephone service, and even CATV, but the results have been mixed. In many ways the Babies were learning about shaping their destiny and were flexing their muscles. The Baby Bells now seem to have entered their rebellious teen years and believe anything is possible. And like teenagers, they want no restrictions on what they can do.

Over the past few years, the Baby Bells have mounted a well coordinated, collective campaign to be regulated less at the local level and to be allowed into manufacturing, long-distance, video entertainment, or anything else they wish to do. The Baby Bells financed studies that concluded that the telecommunication infrastructure in the United States had fallen behind most other industrial countries. The only way to remedy the situation, the Baby Bells claimed, was to free them from regulation so that they could invest in the new technology to construct a new infrastructure. A vision of a telecommunication Utopia was dangled before politicians along with promises of revolutions in the delivery of education and other social services. Of course, this campaign was hyperbole, but the Baby Bells achieved their goal of less regulation. By dangling large purchases of equipment before AT&T, the Baby Bells effectively neutralized the one opponent that might have been strong enough to block their way.

The one force blocking their way was the MFJ and its restrictions. The Baby Bells promised that if they were allowed into cable television, they would rewire the Nation with a new broadband network that would allow the delivery of video and a host of other new services. By allowing the Baby Bells and other local telephone companies to compete with CATV firms, it was hoped that consumers would see lower charges, and that less regulation of CATV would be needed. CATV firms would be able to offer telephone service and thus less, or perhaps even no, regulation of them would be needed.

The vision was that we would have telephone companies competing with CATV companies, and different telephone companies competing with each other in the provision of CATV and telephone services. Regulation would no longer be needed, and competitive market forces would shape the future, without government involvement.

All this looks great—after all, who is not in favor of competition. But as we saw earlier, the technology to allow all this competition simply does not exist. Furthermore, the collapsed Bell Atlantic/TCI merger shows that the Baby Bells will most likely use their financial power simply to purchase CATV firms, and there might actually be less competition. Or, the Baby Bells will use their great financial wealth to over-build cable and then undercut the existing CATV providers to force them into bankruptcy.

As we saw in the previous chapter, the Baby Bells are becoming increasingly intertwined and are acting more and more like a single entity. In fact, Pacific Telesis will be acquired by SBC, and Bell Atlantic will acquire NYNEX. The telecommunication superhighway could become a road leading to greater control in the hands of a few large firms.

Perhaps the past central direction of AT&T kept the local telephone companies under control. Now that the Baby Bells are free of any parental control, they are testing the waters to see exactly how far they can go. They approached the United States Congress in an attempt to bypass Judge Greene and obtain legislative relief from the restrictions of the MFJ—and they succeeded!

THE CONGRESS REWRITES TELECOMMUNICATION POLICY

Early in 1996, the United States Congress overwhelmingly passed—and President Clinton signed—the Telecommunication Act of 1996. This legislation is intended to promote more competition in telecommunication by allowing telephone companies to provide CATV service and CATV firms to provide telephone service.

This kind of attempt to reform telecommunication through

legislation is not new. In 1979, similar attempts were made to reformulate the telecommunication industry through legislation. Legislation proposed then in the Senate by Senators Ernest Hollings and Barry Goldwater would have allowed the telephone companies to provide video transmission but not program content. Senator Goldwater's legislation would have prohibited the CATV firms from providing telephone service, while Senator Hollings' legislation would have allowed the CATV firms to provide telephone service, subject to regulation as common carriers. In the House, legislation proposed by Representative Van Deerlin would have allowed the telephone company to provide both video transmission and content and the CATV firm to provide all telecommunication services including telephone service. All these legislative initiatives of the late 1970s failed, perhaps because of lobbying pressures against them, and ultimately, because of lack of interest.

Today's legislative attempt did not fail. Somehow a compromise was found to satisfy the conflicting views of all the various parties. More pressing affairs—such as the federal deficit or a presidential election—did not distract the attention of the legislators. Or, perhaps all these more pressing affairs so distracted the attention of the legislators that they passed legislation just to make the lobbyists happy. Whatever the reasons for its passage, it was without much oversight or analysis of the consequences.

In its attempt to eliminate competition, AT&T embraced the Consumer Communications Reform Act (CCRA) of 1976, which would have virtually outlawed competition and protected the Bell monopoly. This attempt failed. The recent lobbying efforts by the Baby Bells for legislative relief from business restrictions is no less self-serving than the CCRA. But this time Congress did not see through all the lobbying. I believe that in the end the new legislation will need to be redone to eliminate all the harm to consumers. Furthermore, all the new rules in the name of competition will only create more conflict. In the end, I would not be surprised to see the Baby Bells lobby against the new legislation.

Undoubtedly, competition is the best way to "regulate" an industry. But the nature of much telecommunication seems to be a natural monopoly, or if not a monopoly, then at least so large in concentration and impact that some form of government regulation is needed. Is there some way to understand the real issues and suggest some framework to set a balance between regulation and competition for the future of telecommunication? I believe the distinction between content and conduit might be an appropriate framework to begin a discussion to unravel some issues.

CONTENT AND CONDUIT

The conduit is the transmission and switching network that carries various telecommunication signals. The content is the telephone, data, and video signals carried over these networks. We saw earlier, that in terms of revenue, content is key in the entertainment industry but conduit is key in the telecommunication industry. Telephone companies are common carriers that must offer service on an equal basis to everyone and must not exercise any control over the content of the signals they carry. On the other hand, CATV operators control over both the conduit and the content. CATV operators are not common carriers.

An interesting idea is to apply the concept of common carriage to the provision of CATV service. The operator would only provide the conduit. Users could access, or dial up, any desired video program from a myriad of video program providers. Thus, the concept of video dial tone, or video on demand, looks like a good idea and would result in more freedom of choice and less control over programming by the provider. However, as we saw in an earlier chapter, the technology to do this is still not at hand, and if it were, would be very costly.

The key question is who would own the conduit for the provision of video entertainment. Currently, CATV operators and local telephone companies seem well positioned in the provision of telecommunication and in theory might compete with each other. However, from an operating efficiency basis, such competition would require duplicated local facilities, which would be costly and inefficient. Hence, it might be more efficient if one of them owned and controlled the conduit and the other packaged and offered the video content. But given recent history and the tremendous financial strength of the Baby Bells and other local telephone companies, it seems most likely that the telephone companies would attempt to purchase CATV firms and in this way control both the conduit and the content. The only way such a concentration of power could be prevented would be through appropriate legislation and restrictions on the local telephone companies to keep them out of the provision of content. Only then would the provision of conduit on a common-carrier basis seem appropriate for them.

A telephone company leasing a cable system to a CATV operator is not new, as we shall see in the next section of this chapter. The real issues center on the provision of programming.

AT&T AND CATV

AT&T and the old Bell System always had an interest in CATV, and in the mid-1960s AT&T actually owned about 75 CATV systems in the United States. These CATV systems were leased to the local cable operator, and AT&T had no involvement in determining the content of programming provided by the cable operator. In 1969, AT&T reviewed its policy regarding ownership of CATV systems and for a variety of reasons sold all but one of the systems during the 1970s. The last system to be sold was in Manhattan, where the cable operator was having trouble obtaining the required franchise from the city.

Perhaps the single most important reason AT&T abandoned the CATV business was that the revenues generated were not favorable compared to the costs. But if the tariffs were accordingly raised, the systems would not be competitive. There also were concerns expressed by the CATV industry about AT&T's presence in the provision of CATV service, and AT&T was encountering delays in obtaining the required authorizations from the FCC.

In 1979 while I was at AT&T, I prepared a presentation on CATV that my management made to AT&T's Chairman of the Board. In the presentation, I predicted a confluence of telephone service and television that would occur in the early 1990s from the progression of technology. This progression would ultimately, in my view then, enable the CATV industry to provide telephone service and the telephone industry to provide television. I concluded that to the extent that the CATV operator could provide telephone service, this was a threat to AT&T. But to the extent that the new technology would enable the telephone company to carry broadcast television in addition to voice and data, the Bell System had a new opportunity.

Today, as a result of the Telecommunications Act of 1996, the Baby Bells are allowed to provide long-distance service, but AT&T and the other long-distance companies are strongly opposed. The storm clouds are gathering for a major battle between AT&T and the Baby Bells.

One appropriate strategic response by AT&T would be to enter the local business in competition with the Baby Bells. Since both the profit margins and total revenues are far greater for the provision of local service than long distance, the Baby Bells have far more to lose from local competition than AT&T has to lose from additional long-distance competition.

One way for AT&T to enter the local area is to develop the technology to allow local telephone service to be provided economically and simply over the existing coaxial cable of local CATV firms. AT&T could sell the technology to CATV firms or enter into strategic partnerships with existing CATV firms. By bypassing the Baby Bells and other local telephone companies, AT&T would avoid payment of access charges and could thus offer lower rates for long-distance service.

Another way AT&T could compete directly with the Baby Bells in the provision of local service would be through radio technology using fixed radio equipment in the home that would bypass the local loop of the Baby Bells. A third way for AT&T to enter the local-service market would be to acquire GTE, since GTE already has a national presence at the local level.

Meanwhile, in order to take on AT&T, the Baby Bells are merging to become bigger—and better, they hope. By the end of 1996, SBC hopes to acquire Pacific Telesis, and Bell Atlantic hopes to acquire NYNEX. Will the US WEST Communications Group join the SBC/PacTel combine? What Bell South join the Bell Atlantic/NYNEX combine? What will Ameritech do to become better by becoming bigger?

The coming AT&T/Baby Bell battle should be exciting indeed and will add to the current chaos afflicting telecommunication in the United States.

VIDEO DIAL TONE

Much earlier in this book we saw that the CATV industry initiated a threat against the telephone companies 25 years ago by promising to provide telephone service over the CATV coaxial cable. In the end nothing ever happened, for a variety of reasons involving technology, policy, economics, business, and the consumer.

Today, the telephone companies have initiated a threat against the CATV companies by stating their intentions to provide television programs that are dialed up over the telephone network, or over some new infrastructure owned by the telephone companies. This new approach to the provision of video entertainment is called video dial tone. Some CATV companies have attempted to counter this threat by investigating ways to offer telephone service over CATV facilities.

The FCC's position is to foster competition, and accordingly the FCC approved applications by telephone companies to offer cable service. The FCC approved a trial of video dial tone, to be conducted by Bell South in Georgia, offering video only over a system of 70 analog and 240 digital channels. The FCC also approved a request by NYNEX to build commercial video dial tone systems. The FCC gave approval to Ameritech to build a hybrid fiber/coax system for video dial tone with 240 digital, 70 analog, and 80 switched video channels. Ameritech predicted 1.3 million users by the end of 1995 and 6 million users by 2000. Ameritech plans to own and provide programming. Ameritech is also very active in over building cable systems using present coaxial cable technology.

The 1984 Cable Act forbad telephone companies to provide video programming in their own service areas. Citing their First Amendment rights to free speech, the telephone companies were successful in convincing the courts to overturn these restrictions. I can well understand how the First Amendment protects the free speech of Baby Bell executives and also the commercial speech of the companies themselves. But video programming is a product, and why First Amendment should protect the right of the Baby Bells to provide this specific product is not that clear to me. The Telecommunications Act of 1996 ends this debate by allowing the Baby Bells and other local telephone companies to provide CATV service.

CABLE MODELS

There are a number of models for the provision of video programming. The conventional cable model is that the operator of the cable system also be responsible for the selection and provision of all programming. The common carrier model is that the operator of the cable system have no control over programming, not provide any programming, and nonpreferentially carry any and all programming provided by others .

The Telecommunications Act of 1996 created a new type of cable service—called Open Video Systems (OVS)—for local telephone companies. With OVS, the operator of the cable service can provide programming but must also open the system to other programmers on a nonpreferential basis. An OVS operator will be subject to less regulation but in return must not discriminate among video providers carried over the OVS. If demand exceeds the channel capacity of the OVS, then the OVS operator's control is restricted to one-third the capacity. The FCC is given the task of specifying detailed rules for OVS before the end of 1996.

VERTICAL INTEGRATION

The old Bell System was a vertically integrated monopoly. Bell owned everything from the telephone instruments and wiring in homes to the long distance cables across the country. Today, consumers own and provide their telephones, facsimile machines, and modems that are connected to the telephone outlet. Consumers also own the telephone wires in their homes, although they can contract with the local telephone company to service this wiring.

Perhaps at one time in the distant past it made sense for the telephone company to own and provide telephones and inside wiring. But today it does not make any more sense to rent a $20 telephone from the telephone company, than it would to rent an electric toaster from the electric company. But years ago, the electric company owned the inside wiring, outlets, switches, and even the light bulbs. Once this vertical integration of electricity was broken, a vigorous electric appliance industry formed with inventions of all sorts of new appliances.

During the days of telephone ownership by the telephone companies, it was illegal to own and connect a non-Bell telephone to the network. I have heard stories of Bell employees entering people's homes and removing any non-Bell phones. It was even forbidden to attach a neck rest to a Bell telephone handset, or even a plastic cover to a Bell telephone directory. Thankfully, those days are gone—we have removed the phone company from our homes.

It was very costly to send a telephone repair person to a home to repair a $20 telephone. Hence, telephone companies should be thankful that they are no longer in the business of intruding into people's homes. It is far less trouble to provide telephone service to the customer's premises, and not to be responsible for what happens inside the premises.

Today, the cable company owns the set-top convertor box and the inside wiring. The cable company has taken over the role of intruder into homes. When the local telephone companies provide video they too will be back in homes with ownership of set-top convertor boxes and inside wiring. Is this what consumers really want? Do the telephone companies really want to be back into this very labor-intensive business?

UNIVERSAL SERVICE

Telephone service is only valuable if everyone has it. Alexander Graham Bell, the telephone's inventor, and Theodore N. Vail, the Bell System's creator, both recognized this fact. Thus, the concept of affordable, universal telephone service was born. The Communications Act of 1934 formally endorsed universal service and made it national policy. It is interesting that sixty years later, universal service—its definition and how to achieve it—is still an issue.

Universal telephone service has been accepted as nothing more than dial tone and access to the telephone network. Touchtone, caller-ID, voice-mail, and other such services are all beyond basic universal service. But now that video dial tone has been proposed, the question arises whether it should be included as part of universal service. Broadband service has been proposed also, and the same question arises about whether it should be part of universal service.

Video dial tone and broadband service are very costly and probably would only be affordable by a small number of households initially. Thus, telephone companies do not want to be forced to make these services available to everyone as part of universal service. But national policy does not want to create a class of citizens who do not have equal access to all the information that supposedly would come over the superhighway of broadband and of video dial tone. The same phone companies who promoted video dial tone and broadband service now find themselves arguing against providing these services to everyone.

DIVESTITURE CONTINUES

AT&T divested itself of the local telephone companies in 1984. Now some of the Baby Bells are continuing the process of divestiture. A couple of years ago, Pacific Telesis surprised everyone with the announcement that is was considering divesting itself of its basic telephone business. I did a quick back-of-the-envelope financial analysis and concluded that such a divestiture would not make much financial sense, since the provision of telephone service was the cash cow, and cellular and all the new ventures were not nearly as profitable.

In the end, reason prevailed, and Pacific Telesis instead spun off its cellular and other new ventures in 1994 to create a new company, AirTouch. In 1995, US WEST made a similar move by creating two classes of stock: one for the telephone business, the US WEST Communications Group, and the other for wireless and other new business ventures, the US WEST Media Group. Also in 1995, AT&T announced its plans to divest its manufacturing and computer businesses. Thus the process of divestiture continues.

We saw earlier that the financial profiles of a telephone company are quite different than most media firms. Breaking up a Baby Bell, either directly through divestiture or through separate stock, acknowledges these differences and the different types of investors that are attracted to both. It also makes it much easier to see how risky the new ventures really are and how well, or poorly, they are doing. It also might make it easier for the United States Department of Justice to force a breakup on the Baby Bells, if such a remedy is ever needed.

Like CATV companies, the local telephone companies are monopolists at heart. We all know that monopoly is evil and must be destroyed in favor of competition. Isn't this why we broke up the old Bell System? Yet monopoly with effective regulation might very well be an appropriate model for the provision of local telephone and of CATV services. The action of the telephone companies might be to extend their local monopoly to include the provision of entertainment and long-distance service by the outright acquisition of CATV firms and long-distance companies. If so, then the old Bell System might reform right before our eyes, but with a far larger monopoly than ever.

The plans of SBC to acquire Pacific Telesis and of Bell Atlantic to acquire NYNEX indicate that such a reformation is indeed quite possible. But would such a reformation be acceptable to the Department of Justice, and if so, what safeguards would be needed? Would the seven Baby Bells be allowed to merge and then purchase AT&T to provide long-distance service?

In 1974 people were convinced that I was crazy when I suggested that by 1984 the local Bell companies would be independent of AT&T. Yet this divestiture occurred exactly as I predicted. AT&T said I was wrong when in 1991 I suggested that AT&T should divest itself of all manufacturing. Yet this divestiture will occur by 1997. It seems that crazy things are normal in today's telecommunication industry, and the crazier the prediction, the more likely it seems that it will occur.

THE ANPA VERSUS AT&T

For decades, the newspaper industry has been fearful of electronic yellow-page services that might be offered by AT&T and the phone companies. Such electronic services would threaten the newspaper's lucrative classified ads.

About 15 years ago, the AT&T directory organization announced its intention to conduct a trial of an electronic Yellow-Pages service in Austin, Texas. The American Newspaper Publishers Association (ANPA) was able to obtain a court injunction against the trial and halt AT&T's plans. However, in the same time frame, a different part of AT&T was jointly conducting a trial of an electronic newspaper service with Knight-Ridder Newspapers in southern Florida, thereby demonstrating quite clearly that AT&T had no plan to dominate electronic publishing.

At the time of the Judge Green's review of the MFJ, the ANPA was able to obtain restrictions in the MFJ to prevent AT&T from offering any electronic publishing over its own facilities for a period of seven years from the 1984 divestiture. The Telecommunications Act of 1996 prohibits the Baby Bells from offering electronic publishing for four years from the date of enactment.

STANDARDS

Standards are very important. However, standards are sometimes overemphasized. Consumers buy a specific product or service—not a standard. An example of this occurred with AM stereo. The standards battle over a number of different approaches to AM stereo was quite fierce. The Federal Communications Commission decided that it would not decide on a standard and that instead the marketplace and radio industry should determine which standard was best. In the end, it did not really matter what standard was chosen, as there was no real market need for AM stereo. AM broadcast radio is "lo-fi" as contrasted with the "hi-fi" of FM broadcast radio. Stereo indeed makes sense for FM, but little sense for AM.

Standards can actually impede progress. Once a standard has been chosen, further technological advancement might make obsolete the chosen standard. The progression of standards for HDTV (High Definition TV) is a good example of this. The original approach to HDTV used analog technology. Within a few years, a digital approach using compression was suggested, and a number of different approaches have appeared. Given the continued fast pace of technological advances in compression, any chosen standard will most likely become obsolete.

TV REGULATION

The publication of newspapers is not regulated. The "freedom of the press" provisions of the Constitution of the United States prohibit such control. Television is the means by which many people obtain news, but since TV uses the scarce resource of the public airwaves, it is subject to regulation. To prevent a broadcasting monopoly, the FCC sets rules on the number of TV stations that can be owned by one broadcaster. These rules have been liberalized and changed from time to time and were even more liberalized by the Telecommunications Act of 1996.

One major rationale for regulation is scarcity and lack of competition. Today, we can obtain video programs from a large variety of sources: over-the-air VHF and UHF broadcast television, direct broadcast satellite, video tapes purchased and rented from video stores, video disks, and CATV. Given all these sources, is there a lack of competition?

CATV has not made any significant penetration in the United Kingdom where over-the-air broadcasting signals are available nearly everywhere and VCR machines are plentiful.

Granted that the local CATV firm has a monopoly on the delivery of video over coaxial cable—controlling both the conduit and the content—does this constitute a true monopoly given all the other sources of video programming in the United States? But all the philosophical arguments mean little when the public becomes angered over CATV rate increases and clamors for re-regulation.

CATV started as a "mom-and-pop" business bringing TV to areas not otherwise touched by the radio signals of television. Thus CATV was originally ignored by the regulators, since it was perceived as a good thing. As a result of pressure from local TV stations, the FCC reluctantly initiated regulation of CATV in 1965, with rules on the importation of distant TV signals—the intent being to protect the local stations from competition.

Over the years, regulation of CATV has increased—and decreased—depending upon the political climate and has been extended to rates. For example, rate regulation of CATV was initiated in 1988 and then terminated a few years later only to be initiated again in response to congressional pressure. The Telecommunications Act of 1996 deregulated CATV again. CATV has had a history of regulation, de-regulation, and re-regulation. I imagine that this oscillation will continue.

TELEPHONE REGULATION

At the time of the Bell breakup, AT&T assumed it would no longer be regulated after the breakup, since competition was established in the provision of long-distance service. Today, well over ten years since the breakup, AT&T's long-distance business is still regulated by the FCC, under the justification that AT&T is the dominant carrier, even though AT&T's market share has fallen to only about 60 percent.

The Baby Bells are also still regulated, which seems understandable since at the local level they still have a monopoly on the provision of local telephone service. But cellular service competes with wired local service, which offers the possibility for some form of competition at the local level based on transmission medium: wireless versus wireled.

Before the Bell breakup, long distance calls within a state were carried by the local company which retained all the revenue. After the breakup, the Baby Bells were forbidden to carry this intrastate long-distance traffic and were restricted to providing telephone service to only their local access and transport areas (LATAs).

During the past year or so, long-distance companies have started to compete with the local telephone companies in providing toll service within LATAs. The local companies have decreased their toll rates in an attempt to compete and have offset this potential loss in revenue by increases in the rates for basic service. However, the more they charge for local basic service, the more lucrative it becomes for the CATV company or someone else to compete with the local telephone company in this area. But, as we learned in an earlier chapter, the technology that would allow a CATV company to provide telephone service over the coaxial cable is still not at hand at competitive costs. We shall simply have to wait and see what the future brings.

SUBSIDIZATION

The goal of universal telephone service was to be achieved by making basic telephone service inexpensive and affordable to everyone. To achieve this goal, some customers and classes of service were charged much more than cost to create a subsidy to keep basic service inexpensive. Thus, long distance rates were kept artificially high to create profits that would subsidize local rates for basic telephone service. Business customers were charged about twice as much for basic service as residential customers, again to create a subsidy. On thing that has been learned from the Bell breakup is that subsidizations create opportunities for competitors who do not have to subsidize anything. Thus, MCI initially was able to "compete" with AT&T since MCI did not have to subsidize local service.

There is some controversy in my mind about subsidization since the actual true costs of providing telephone service for different classes of customers is difficult to determine. Hence, although it is generally accepted that business subsidizes residential service, I have never seen any real studies to support this claim. Since the residential telephone is used much less than the average business phone, one would expect businesses to pay more for basic service.

9-1-1 is the universal telephone number for an emergency. Telephone companies are required by law to collect a small sum from telephone subscribers to subsidize the costs of 9-1-1 service. However, police and emergency organizations have become innovative in extending the definition of 9-1-1 to include virtually all aspects of emergency communication, even to radios in police cars and to cellular phones. Once a subsidization is initiated it is almost impossible to remove. Also, the telephone bill can become a new source of "taxes" to subsidize social projects.

In New Jersey, very low rates are offered to educators to connect classrooms with two-way video. Once initiated by Bell Atlantic, this is a form of subsidization that will be difficult to halt. The Telecommunications Act of 1996 actually mandates lower rates for schools and libraries. The telephone company should not, in my opinion, be in the position of determining societal priorities and then subsidizing certain customers or projects.

THE NEVER-ENDING POLICY PROCESS

The policy uncertainties continue as this book is being written. The right of free speech, the freedom of the press, cross-ownership rules, concentration of power and media ownership, and the distinctions between content and conduit are just a small sample of the considerations that are affecting the never-ending policy debate.

Policy in the United States is usually not determined on some broad, overreaching, intellectual basis, but rather evolves from many meandering decisions, false starts, and even drastic changes along the way. Today's apparent winner can quickly become tomorrow's loser. The fun of policy is that nearly anyone can have an opinion and their opinion is usually as expert as anyone else's. The policy debate about regulation, business restrictions, video dial tone, and other such issues are most certainty not over—the Telecommunications Act of 1996 notwithstanding. So continue to stay tuned to the policy channel—more is to come.

READINGS

Arnst, Catherine, "I-Way Tie-Up on Capitol Hill," *Business Week*, February 6, 1995, pp. 146-147.

Cauley, Leslie, "Seven Baby Bells Win Right to Provide Long-Distance to Cellular Customers," *The Wall Street Journal*, May 1, 1995, p. B4.

Noam, Eli, "From the Network of Networks to the System of Systems," *Regulation*, 1993, Number 2, pp. 26-33.

Van, Jon, "Merger may lead to a Bell family reunion," *Chicago Tribune*, April 2,1996, pp. 1 & 12.

"Gore's law," *The Economist*, January 15, 1994, p. 72.

"Taking the scenic route," *The Economist*, April 16, 1994, pp. 67-68, 71.

"The movie mogul on the line," *The Economist*, May 22, 1993, pp. 67-68.

"Vice President Proposes National Telecommunication Reform," White House Press Release, January 11, 1994.

"Washington's wake-up call," *The Economist*, January 20, 1996, pp. 61-63.

Chapter 10

THE INTERNET EXPOSED

If any one service is central to the superhighway, it is the Internet. The Internet is a global network of networks spanning the world; linking the large computer systems of universities, government, and industry; and enabling users to send text messages to each other and to interchange other forms of digital data.

It has been predicted that the Internet will have 200 million users by 2000. We shall see whether this turns out to be just another example of hope and wishful thinking. The basic idea of a universal form of text communication over an appropriate, public data network is appealing—although the idea of telecommunication through text is as old as the telegraph.

Data networking has been adopted and glamorized by Vice President Gore, and as a result, the Internet is at the heart of the Clinton administration's vision of a National Information Infrastructure (NII). Thus, the Internet and data networking have become national policy under the Clinton administration. The Telecommunications Act of 1996 makes it the policy of the United States to promote "the Internet and other interactive computer services and other interactive media." However, packet switched data networking and e-mail are more than two decades old. Similarly, electronic access to data bases along with remote browsing of library holdings and other such text-based information services are most certainly not new. We know that users will use these services; the real issue is their cost.

This chapter explores the Internet and some of the issues that surround it and packet switching.

INTERNET TECHNOLOGY AND HISTORY

The core of the Internet is a world-wide, packet-switched network for data communication. Packet-switching is efficient and appropriate for the bursty, short messages that comprise most data traffic. The data traffic is organized into packets, with each packet consisting of a few thousand bits of data. A digital address of the destination is affixed as a header to each packet. The address of each packet is examined at switching nodes, and the packet is then sent along to the next node closest to the destination. The actual path depends upon the traffic being carried between the various nodes, and if a specific path is busy, an alternative path will be used. The Internet is composed of many data networks around the world is thus is more of a virtual network than a single physical network.

The ARPANET, developed by the Advanced Research Projects Agency of the Defense Department, was the first packet-switched network. The ARPANET initiated service in the late 1960s and in the late 1980s became assumed by the NSFNET, which was supported and operated by the National Science Foundation (NSF). The NSFNET consisted of a backbone packet-switched network, called National Backbone Service and supported financially by the NSF, and about a dozen regional packet-switched networks, also supported by the NSF. Although a number of telecommunication firms offered their own packet-switched networks, the NSFNET became the backbone packet-switched network for the Internet in the United States.

The NSFNET started in the mid 1980s. The backbone packet-switched network then was formed from data lines operating at 56 kbps (known technically as DS-0 lines). In 1987, the data lines in the backbone packet-switched network were increased in capacity to 1.54 Mbps (known technically as DS-1, or T1, lines). Again, in 1992, the capacity of the backbone lines had to be increased to carry all the traffic, this time to 45 Mbps (known technically as DS-3, or T3, lines). The backbone network is supplied by MCI as a virtual network of T3 lines. IBM supplies the routers that examine the destinations of the packets and then switch the packets on their way along the network.

The data traffic carried on the Internet is organized into packets with each packet consisting of a few thousand bits of data. A byte is 8 bits of data and can correspond to a single alphanumeric character. Most of the traffic carried over the Internet is large files of information, called file-transfer-protocol packets (FTP), which swamp e-mail traffic by a factor of 40-to-1. The Internet is reported to have carried 20 billion file-transfer-protocol packets and 0.5 billion e-mail packets in September 1994. About 80 percent of the addresses on the Internet belong to commercial firms —indicated by the suffix ".com" in the address.

The Internet was reported to have over 20 million users in 1994. The 20 billion packets, carried in September 1994, thus are equivalent to an average user sending about 1,000 packets per month, or about 30 packets a day. This is very little average usage. Assuming an average packet size of 500 bytes, or about 4,000 bits of data, this average usage is equivalent to only 2 seconds of digital speech. Packet switching is a very efficient way to transmit data and is relatively inexpensive.

In 1994, the NSF spent $26 million to support NSFNET: $14 million to $16 million for the backbone, $8 million to $10 million for regional networks, and $2M for connect fees for user institutions to gain local access through their local telephone companies. Not surprisingly, since the amount of traffic is minuscule compared to the telephone network, the NSFNET does not cost that much to operate, particularly when compared to the $200 billion of revenue generated by the entire telecommunication industry. In addition to the $26 million per year spent by the NSF, other federal agencies are also contributing to the support of the Internet. The NSF is in the process of ceasing its support of the NSFNET. In 1995, the NSF ceased its support of the backbone, and over the next four years will cease its support of the regional networks.

The NSFNET and the Internet were originally developed by computer scientists for use by other computer scientists and, as you might expect, were not easy to use by mere mortals. Friendly software interfaces were developed, such as Mosaic. These interfaces utilize pull-down menus and graphics, just like Macs and Windows, to make access more friendly and easier to nonspecialists. The World Wide Web is a multimedia presentation of information with friendly links from one database to another.

PACKET SWITCHING

The packet switching used by the Internet is not a new concept. Packet switching was invented in the 1960s. Unlike circuit switching in which a complete connection is maintained between two parties for the duration of the communication, packet switching transmits and routes short packets of data between computers. Packet switching is most appropriate for the short bursts of data that people send to a computer and the messages sent back.

I first learned about packet switching in the early 1970s when I went to Washington to work on the staff of the White House's Science Advisor. The ARPANET, the first packet-switched network, had been developed and operated by the Advanced Research Projects Agency (ARPA) of the Defense Department to link together the large computers used by the researchers supported by ARPA. It was believed that the researchers would use the ARPANET to share computing resources and to exchange computer programs. Instead, the researchers used the ARPANET to exchange memos, text messages, and documents—perhaps the first use of e-mail. Again, we see that users are more inventive in discovering uses for technology than the engineers.

The ARPANET leased transmission lines from common carriers. The backbone network consisted of single 56 kbps line which was sufficient to carry all the traffic. Thus, although there was much hoopla over the ARPANET, it actually was an extremely thin network in terms of transmission capacity. The same is true of today's Internet.

Given all the hype over the Internet, along with the confusion over exactly what it is, I wonder whether we should hype the telephone network similarly. After all, by using the telephone I can reach virtually anyone anywhere at any time to obtain virtually any information about virtually anything. Perhaps we should rename the telephone network "HumaNet," since I can reach any human!

A great debate was raging in the early 1970s over whether the use of the ARPANET—since it linked computers—was computing or communication. I was not puzzled by this debate, since in my mind the ARPANET was a communication network used for communication purposes. The fact that the content of the communication could be computer access did not make the ARPANET a computing function. The question had also become clouded since the ARPANET used routers to switch the packets along the way, and these routers were digital computers. Again, the fact that digital computers were used as packet switches did not change the overall purpose of ARPANET, namely, communication. This is similar to today's Internet. The Internet really consists of two separate functions: one is a worldwide packet-switched network; the other is all the various data bases and information that can be accessed over the packet-switched network.

Another debate in the early 1970s involving the ARPANET centered around its use by non-ARPA researchers. In particular, researchers supported by the National Science Foundation wanted access to the ARPANET. The question was how to give them this access. ARPA was not licensed as a common carrier and hence could not sell access. The NSF researchers were not ARPA contractors and thus could not be given free access.

I remember a meeting held in 1972 by the White House Office of Telecommunications Policy (OTP) to confront ARPA with these facts. I attended at the request of my boss, the Science Advisor. ARPA was sternly warned that it should not sell access to the ARPANET. I questioned whether a solution would be for private industry to offer a commercial version of the ARPANET and whether AT&T would be one company that might wish to offer such a service. I was told that AT&T had been asked but had no interest in offering commercial packet switching. I then asked whether ARPANET's prime contractor Bolt, Beranek, & Newman, Inc. (BBN) would be interested, particularly since they were already so heavily involved in the project. I was told that they too had been asked and were not interested.

A few days later, I was talking by phone with Dick Bolt, the "Bolt" of BBN, and told him what had transpired at the OTP meeting. He seemed surprised and said he would look into the matter. A few days later, he called me to tell me that someone at BBN had made a poor decision and that now BBN was very much interested and would soon start a new company to offer packet-switching. That company was Telenet.

Private financing was obtained to create Telenet, and the private placement was completed in 1974. In 1979, GTE purchased Telenet. Southern Pacific Communications was founded in 1983 to provide long-distance service in competition with AT&T using the right-of-way along its railroads. Southern Pacific Communications became US Sprint and was acquired in 1986 by GTE. Telenet was then moved into Sprint, and when GTE divested Sprint, Telenet went along with the divestiture and is now called Sprint International. What this means is that a packet-switched network is already available commercially in the United States. The only rationale for NSFNET is simply as another form of federal subsidy of universities and university researchers. But NSF has a history of such subsidization in the name of research.

I remember that a big issue in the early 1970s was the NSF subsidization of science information services. After much bickering between the NSF and the White House Office of Management and Budget (OMB), the NSF finally decided to halt the subsidization and to allow commercial market factors to decide the future of such information services. Somehow, the NSF got out of one area of subsidization and fell into another as it expanded into the NSFNET and the Internet.

FEDERAL ROLE

The NSF justified its support of the NSFNET as "priming the pump" to encourage electronic communication among the researchers financed by the NSF. However, these researchers also use the telephone and facsimile to communicate among themselves. Should the NSF therefore create and support a special national research and education network to carry telephone and fax calls?

Packet switching and data communication are not new concepts and have been available commercially for decades. Packet switched data communication is available commercially today from telephone companies and long-distance carriers.

I do not accept NSF's subsidization of the data communication of university researchers and practically anyone else using the Internet. I am disturbed that the NSF got itself into the position of being a telecommunication carrier in competition with private industry. This is clearly in violation of policies against federal competition with private industry. But I guess if I could invent a way to get the federal government to pay my personal phone bill in the name of research and scientific communication, I would do so.

The issues surrounding the Internet are far more extensive than just the NSF discovering another way to subsidize universities. Much of the traffic carried on the Internet and the NSFNET is commercial, non-university traffic. In October 1994, 16,744 of the Internet addresses were commercial organizations and only 4,140 addresses belonged to education, government, and nonprofit organizations. In other words, 80 percent of Internet addresses were commercial organizations. This tells me that the NSF was using tax dollars to subsidize the data communication of commercial firms.

In many ways, the Internet's problem is that it has been too successful. Nearly anyone can obtain "free" access just by knowing someone who can get you an address and account. Any commercial organization with a computer can obtain direct access and use the Internet to transmit e-mail and data for "free' around the world—all mostly paid for by the NSF and other federal agencies—and ultimately by your and my tax bills. And to make matters worse, there seems to be a general belief that many of the large files sent over the Internet are pornographic images. Information about the construction of bombs and explosives are rumored to circulate over the Internet. Computer nerds have used the Internet to gain access to sensitive government information and to destroy information in data banks.

All these problems create a need for federal regulation of the information being sent over the Internet, but this then raises the question whether such regulation would constitute censorship of free speech. The Communications Decency Act of 1996, signed into law along with the Telecommunications Act of 1996, prohibits the transmission of "patently offensive" material over the Internet to persons under 18 years of age. Many groups are concerned about this apparent violation of free speech.

The use of the Internet to send e-mail messages between scientists and researchers was all for the good, but the Internet exploded beyond this application and became more and more of an embarrassment to the NSF, particularly as the proportion of legitimate research users became smaller and smaller.

Given the growing number of problems and issues surrounding the Internet, the NSF is wise to cease its support of the NSFNET. However, a dole once created is difficult to stop, especially abruptly. Those on the dole come to expect their subsidy as a right and clamor for it to continue. The issue then becomes how to transition all the users being subsidized into the real world. Packet switched networks are available commercially from a number of vendors and a transition to commercial accounting and service would seem straightforward to me. Let the users pay real prices, and then we shall see how useful and valuable packet switching and the Internet really are.

NSF entered packet switching decades ago in the belief that industry was not willing to provide packet-switched networks. Although I think this belief was more myth than reality, it most certainly is not true today, and the NSF subsidization probably continued far longer than was ever justified. All that is now changing as the NSF gradually ceases its financial support of the NSFNET and the Internet. However, given my knowledge of Washington politics, I fear that some other federal agency will step in and continue the support under some new guise.

E-MAIL

While at AT&T, I became a believer in the market for telecommunication of digitized text, what in 1978 I called "digi-text" and what today is called electronic mail, or e-mail for short.

Using rates for commercial packet-switched communication in the late 1970s, I estimated that digi-text would generate very small revenues compared to voice telephone revenues. I also concluded that the sale of home terminals and other communication hardware would outstrip the revenue from packet-data network usage. I still hold similar beliefs.

I would use e-mail as a means to obtain telephone messages, since listening to my voice mail is very time consuming and I could read text-based messages in much less time. However, I would then like a dedicated e-mail terminal. Years ago, I envisioned an e-mail terminal that had an alphanumeric keyboard, a display, and a telephone handset all integrated in a friendly, easy-to-use manner. This terminal would be my communication center and would be able to store messages unattended. It also would contain my electronic business card with my name and address that could be sent at will as a short burst of data during a telephone call. The French Minitel already exists and offers much of what I would want in a simple terminal. But in the United States we have become overly fascinated with sophisticated graphics, color, and other costly frills that keep an inexpensive, simple e-mail terminal beyond reach.

While I was at AT&T, I installed a telewriter system between my desk and the departmental secretary's desk. The telewriter is older than the telephone and enables a handwritten message to be sent to a distant location where it appears exactly as it is being written. When the secretary answered my phone, either because I was out of the office or talking on my phone, the secretary wrote the message on the telewriter system. The message appeared on a roll of paper on the telewriter receiver in my office. I did not have to stop by the secretary's desk to check messages, and if I was talking on the phone, the message would appear so that I could decide whether to terminate my existing call for the new one.

My telewriter was very useful to me and taught me the value of simple, existing technology to improve productivity. It also taught me the value of electronic messaging.

I also experimented with computer terminals to allow me to send text messages to a friend in Florida. We could also communicate in real time with the terminals, but this use of text for interactive communication was so frustrating that we frequently picked up the telephone to talk directly. Text seems great for stored messages, but voice is better for interaction and immediate response.

In 1979, I suggested that AT&T develop a shared data line approach to enable its customers to have easy access to packet switching without tying up the regular telephone line. My idea was that hundreds of homes would all share a single twisted pair. The line would carry a digital bit stream operating at a medium rate of the few 100 kbps. Since digital text and e-mail are short and bursty, each home would on the average need a very low bit rate. My shared data line, although I most certainly realize that many others had similar ideas, became today's local area networks. I still believe that this kind of system using the local loop of the telephone network could provide a valuable service to many people.

SOME INTERNET FALSE ECONOMICS

Because of the federal subsidy, the Internet has been overly cheap and even free to most users. University users get it totally for free, and non-university users pay very little since they are only marginal-cost users. Of course, all this will change now that the NSF is ceasing its financial support of the NSFNET. MCI offers commercial access to the Internet for a one-time, up-front fee of about $70 and connect charges of about $3 per hour for up to 28.8 kbps access. On-line services such as Prodigy and Compuserve charge similar low rates for access to the Internet. These commercial access charges have been low because of the federal subsidy of the NSFNET, which forms much of the Internet in the United States.

An AT&T service, AT&T Mail, offers e-mail to any address on the Internet or other packet-switched networks for a monthly fee of $3 and rates of 50 cents for the first 1,000 bytes decreasing to 5 cents per kilobyte after the initial 3 kilobytes. Sprint offers a similar service, called SprintNet®, over its packet-switched X.25 network for business-day rates of 23 cents per minute and 5 cents per kilobyte with a monthly minimum of $20. Assuming an Internet flat access charge of $3 per hour for rates of 28.8 kbps, the rate per kilobyte on the Internet comes to about only 0.02 cents. Clearly, the AT&T and Sprint charges are substantially higher than the Internet because of the federal subsidy. It has always been difficult to compete with the federal government and its ability to dip deeply into taxpayer pockets.

INTERNET TELEPHONY

Voice can be digitized and then sent as packets over the Internet or any other packet-switched network. The problem is that packet-switching is not optimum for voice traffic because of the overhead of the packet addresses and the delays that might be encountered in receiving and assembling packets at the destination.

VocalTec, an Israeli firm, sells a software package for compressing speech to about 7,000 bps for transmission over the Internet. With this software, it is claimed that "free" long-distance calls can be made around the world for the cost of only a local call. *The Economist* (February 18, 1995), which usually does better at recognizing hype, warns "tremble, Ma Bell, tremble." But the Internet is designed to carry packet-switched data with its bursty nature—not the continuous and relatively long connect times of voice calls. Furthermore, traffic on the Internet can be delayed, which would be unacceptable for a voice telephone call. And lastly, the Internet is only "free" because of its federal subsidy. Such are the false economics of the Internet.

WHY I'M NOT ON INTERNET

Please do not even try to find my Internet address—I do not have one! Although I am very much a believer in the efficiency of e-mail, I am not on the Internet. Why not? There are a number of reasons.

One reason is that to use the Internet, I would need to connect my Mac to my modem and then to a telephone line and then to the Internet. The Internet is not secure, and all sorts of viruses and software bugs afflict the information that it carries. My Mac is in perfect health, and I will not risk exposing it to the ills that might afflict it and its data by exposure to the public health risks of the Internet. Another reason is that it is too easy to reach people on the Internet. I am already bombarded by too much information and the demands of reading hundreds of e-mail messages would overwhelm me. Those who really want to communicate with me use the telephone or the old-fashioned letter. For the time being,

I will continue to avoid the Internet and cyber-overload.

The owner of the @Cafe in New York City generously allowed the attendees at a conference in 1995 to have free access to the Internet from the computers at each table until 7 PM. I joined a table of artists cruising the World Wide Web. We had a hard time finding anything of much interest and judged the experience boring. The hour of 7 PM finally arrived, and we were told that there would be a charge for any further access of the Internet. Our table and other tables then turned off our computers. But then we could not see our food. We asked to turn on the computers, not to use the Internet, but so that the glow of the computer screen would cast a warm glow on our food! Finally, we had discovered real use for the Internet: a cyber candle!

Although I do not use e-mail, I do use the Internet and the World Wide Web. I use it to obtain information about companies and events in the communication industry. What I have discovered is that the Internet is a great way to obtain press releases and other information about companies without the need to make a telephone call to bother a human. Years ago, we discovered that most people would rather use an automatic teller machine than bother a human teller for routine financial transactions. It seems to me that most people will rather use the Internet than bother a human to obtain routine information.

The Internet is not an entertainment medium; it is an information medium. And most information is in text—not graphics. As an information seeker, I am bothered by the fancy logos and use of graphics at many Internet addresses, or home pages as they are called. The attempts to be creative in the use of graphics simply confuse and frustrate me as I am forced to break the code to find the information I am after. My advice in designing a home page is "keep it simple," minimize the use of fancy graphics, do not overuse color, and emphasize text and a clean, uncluttered layout.

INTERNET BOTTLENECKS

A data communication network, such as the Internet, consists of a number of networks and computer facilities that interconnect to create end-to-end communication. If the traffic exceeds capacity, a bottleneck to the communication can occur at any one of the many switches, transmission facilities, and computers along the way.

Most Internet data communication is asymmetric, with a very low data rate from the user to the source and a much higher data rate from the source to the user. Today's modems operate at 28.8 kbps, which is very close to the theoretical maximum data rate for a 4-kHz telephone connection. Yet many users of the Internet want faster transfers of data. Where are the bottlenecks to faster transmission of data?

Some people have postulated that the bottle neck to faster cyber communication is the local loop with its restricted bandwidth. In fact, the twisted pairs of copper wire of the local loop can carry over 1 Mbps. The central office switch which filters incoming signals to 4 kHZ before digitizing them is the real culprit at the local level.

A solution to the local bottleneck is to use the large bandwidth of the coaxial cable of the CATV system to send data to users. A cable modem would connect user computers to the coaxial cable. A telephone line could be used for the much slower up-stream data signal, or some form of two-way cable technology could be used. Yet another way would be to send short bursts of high-speed data over a direct broadcast satellite channel—a satellite modem.

Much of the need for high-speed data comes from the fancy graphics and use of color that create a friendly, easy-to-use environment. However, the real information is mostly in the text which requires much lower transmission capacity than the graphics and color, all of which are sent over the Internet,. A way to reduce this transmission capacity would be to use local intelligence within the user's personal computer to recreate the friendly interface and send only the real information. In this way, the access speed would be greatly increased and access costs decreased. If the users of the Internet had to pay for the amount of data transmitted, bottlenecks would disappear and terminal-based solutions would be invented.

But even as one bottleneck is overcome, others appear, either within the packet-switched data network or at the various computerized services accessed over the data network. Networks need to be designed from an engineering perspective, examining various alternatives and considering them in terms of their cost and consumer willingness to pay.

TIME-SHARED COMPUTING

Decades ago, digital computers were physically huge and extremely costly. Individual programs were loaded one at a time—a technique called batch processing. Later, this computing power was shared through the use of simple "dumb" terminals—a technique called time-shared computing. The problem with time sharing was that there was insufficient computing power to serve large numbers of simultaneous users and furthermore the software needed for the time sharing itself added much overhead thereby slowing the overall computing response. The ultimate solution was the development of the personal computer. Today's personal computer offers considerably more speed, processing power, memory, and responsiveness at a small fraction of the cost and size of the main-frame computers of the 1960s.

Today, time shared computing has been rediscovered. There are people who suggest that computing be performed centrally and accessed over the Internet from simple, inexpensive terminals. All this sounds much too similar to the time-shared computing of the past. Processing power within personal computers and terminals is here to stay and will not be replaced by central processors. The ARPANET of decades ago was intended as a means of sharing processing power. Instead, users found it a way to share information through e-mail and document transfer. The lessons are to keep processing power as close to the user as possible and to use telecommunication for the exchange of information.

INTERNET REGULATION

Clearly, many people and organizations are using the Internet to move a lot of packet data around the world. What is not as clear is just exactly what data is being sent. The French Teletel system, using the famed Minitel terminal, is supposedly used much for pornographic and sex-chat purposes. Can it be that such similar uses characterize the Internet, as is suspected? Could the transmission of pornographic images account for much of the transfer of large files? Should someone censor the traffic on the Internet? How do we protect children from access to pornographic and other objectionable material? How and who should determine what is objectionable? Do we need new rules and regulations to regulate the Internet and other such packet networks?

The Communications Decency Act of 1996 prohibits the transmission of offensive material. Most Internet users become very concerned when such regulation of content is suggested. And thus we see that like in so many other technology applications, policy and regulation are again an important factor in shaping their future.

And so too is finance an important factor in shaping the future. Some

Internet users make purchases of goods and software over the Internet, and a medium of financial exchange is needed for these purchases. A credit card can be used, but the security of the card number can become an issue. A possible solution is to use digital cash, a form of electronic money that can be used over the Internet for such purchases. The electronic bank of the future dispenses electronic cash and also maintains its security. As one might expect, all sorts of banking policy and regulatory issues arise over this new form of money and financial transactions.

TRIALS

During its early days, the ARPANET was a technology trial with real users. One way to determine whether real users will use and want a new product or service is to offer the product or service in a trial. In the next chapter, we will explore trials and their problems.

READINGS

Karpinski, Richard, "World Wide Web War:Battle of the Browsers," *Interactive Age*, January 16, 1995, pp. 44-46,47.

Lewis, Peter H., "U.S. Begins Privatizing Internet's Operations," *The New York Times*, October 24, 1994, pp. D1, D4.

Memon, Farhan, "I-Phone Has The Baby Bells Beat," *New York Post*, February 23, 1995.

Mills-Scofield, Deborah, "The Internet, from Access to 'Zine," *AT&T Technology*, Vol. 10, No. 3, pp. 2-10.

Noll, A. Michael, "For communication, the Internet is not really where it's @," *The Sunday Star Ledger*, February 25, 1996, Section 10, p. 5

Stoll, Clifford, "The Internet? Bah!" *Newsweek*, February 27, 1995, p. 41.

Turner, Dan, "Internet users having voice conversations," *Los Angeles Business Journal*, May 15, 1995, p. 34.

Verity, John W., "The Internet: How it will change the way you do business," *Business Week*, November 14, 1994, pp. 80-88.

"Facts and friction," *The Economist*, March 2, 1996, p. 72.

"Is there gold in the Internet?" *The Economist*, September 10, 1994, pp. 73-74.

"Tremble, Ma Bell, tremble," *The Economist*, February 18, 1995.

Chapter 11

PITFALLS OF TRIALS

Most people have a difficult time imagining some very novel service if they have had no previous experience with it. Hence, asking people what they think about new services often leads to meaningless results. Most people want to appear positive and embracing toward new technology and therefore will be overly positive in their views. To avoid these biases, a market trial in which real users experience a prototype system seems like an excellent way to ascertain consumer response to new services, such as acceptance and willingness to pay. But many trials are really technical tests of new technology or become blatant publicity stunts.

As will be shown in this chapter, market trials of new services are prone to serious shortcomings and can lead to meaningless results at great cost. In my years at AT&T, I was personally involved in trials of new services in two major areas. One area was two-way, switched visual communication; the second area was videotex and home information services. I will use these personal experiences to draw lessons from the pitfalls of trials.

VIDEOTEX

In the early 1980s, the AT&T marketing department began investigations of the market for a wide variety of new services for residential customers, including remote meter reading, energy management, home security, and home information. The cluster of services was known as Sunray.

AT&T was interested in home information mainly in terms of revenues to be obtained from the sale of home terminals and also from network usage. Another interest was revenue from electronic directories of telephone numbers and electronic Yellow-Page advertisements. AT&T believed in a variety of information services from a variety of information providers. Since the need for home information was not clearly defined, some form of trial with a suitable information provider made good sense. That provider would ultimately be Knight-Ridder Newspapers, then a publisher of 32 daily newspapers, including the *Miami Herald*, the *Philadelphia Enquirer*, and the *San Jose Mercury News*.

AT&T and Knight-Ridder executives met on March 23, 1978 to discuss a possible joint trial. Knight-Ridder Newspapers was concerned about the problems of physical delivery of newspapers, stating that distribution was the single greatest expense and also that the cost of newsprint was continuing to rise. Information delivered electronically over phone lines would solve both of these problems.

A colleague of mine at AT&T and I were given the task of planning the trial and obtaining a commitment to it from AT&T management. I recall that every time I recalculated the costs of the trial, they increased, until my boss finally told me to stop rethinking the costs. The final estimate of AT&T's costs was in the order of $15 million. The final approval of the trial was obtained from AT&T management on April 13, 1978. And thus AT&T and Knight-Ridder entered into a joint trial of videotex.

The trial, called Bowsprit, was conducted in Coral Gables, Florida during 1980. AT&T designed and provided the home terminals, and Southern Bell provided the local network. Knight-Ridder formed a new subsidiary, Viewdata Corporation of America (VCA), to provide the data base and its services.

A few dozen terminals were circulated every four weeks among 125 homes over a six month period. VCA called the service Viewtron. The terminal, which accessed the service over a phone line, had a full alphanumeric keyboard just like a personal computer but used the home TV set for display. Although the British viewdata system used a system of simple block mosaics for graphics, AT&T and VCA were convinced that much fancier graphics were essential and designed a more complex and flexible system, called NAPLPS (for North American Presentation Level Protocol Standard).

Even though the terminals were rotated every few weeks, the

participants all quickly discovered the use of the system to send text messages to the other participants. The use of the system to access information, even though VCA did a spectacular job in creating very pleasing frames of information, was not of much interest to most participants. E-mail was the single largest use. Romances, dating games, finding baby sitters, and simply getting acquainted were big winners of Viewtron's bulletin-board e-mail capability. However, AT&T and Knight-Ridder were convinced that the data base information service, with its advertising and wide variety of data-base providers, was the winning combination. What real users did with the system contradicted the vision of management and simply was ignored. The design of the trial, explained below, also did not actually help the situation.

The participants in the trial were chosen through a process of telephone interviews and focus groups. The participants were finally accepted only if they were positive toward the overall concept of home information. Because positively inclined participants were chosen deliberately, the trial was designed so that only a negative response was meaningful. The participants were treated royally during the trial. With hindsight, given all the positive positioning of the participants, it should have been expected that they responded positively when asked whether they liked Viewtron and that they be willing to pay as much as $30 per month for the service.

Before any real, thorough, independent evaluation could be performed, AT&T and VCA management met on a Knight-Ridder yacht and decided to go ahead with the project and launch an even larger market test followed by a service launch. The project had achieved a life of its own and was unstoppable. The promotion and hype of the trial had became reality.

I remember suggesting that before wasting millions of dollars some form of challenge test be made by offering initial three-month subscriptions to the trial participants for actual money to determine whether they had a real commitment. My management at AT&T was so committed to the project that my suggestion was refused. And thus AT&T and VCA continued to invest heavily in their joint videotex venture. The videotex organizations at both companies grew greatly in personnel.

The service was launched in southern Florida in 1983 with much fanfare and publicity. The terminals and service were sold in large department stores and elsewhere. The promotion and marketing was well done. VCA had signed up many information providers and thus offered a gateway to a wide variety of information services. The only problem was that most consumers could care less. Finally, in 1985 Knight-Ridder admitted defeat and closed down Viewtron. Today, Knight-Ridder admits to having lost $50 million. I estimate that AT&T lost at least twice that amount.

VISUAL COMMUNICATIONS

Rather than admit outright failure with its picturephone in the early 1970s, AT&T mounted an extensive investigation of the market for two-way, switched visual communication. I joined the organization conducting these investigations as an independent, in-house evaluator and critic. AT&T had installed a picturephone system in Bethany-Garfield Hospital in Chicago, and it was receiving good usage. I became suspicious of its claimed success when I visited the hospital and watched people standing before the picturephone units to use them. The camera had a good shot of their belt buckle!

We conducted survey research of the users and discovered that the system was being used primarily as a hot-line telephone system. My management was upset with this finding, since it was negative and upset their promotion of the trial as a success. However, after much discussion, my management finally allowed the publication of a paper reporting the findings of our research. Another application that we investigated was video teleconferencing. Here too the customer response was disappointing.

What I observed was that rather than conducting research we were becoming promoters of the technology. Any open-minded, neutral evaluation was either quite secondary or missing altogether. Promotion of a new service is indeed needed, but the evaluators can not also be the promoters. Any new project and organization will acquire its own life. Independent evaluation will be viewed as a threat to the continued life of the project and organization. A large organization was ultimately created within AT&T to develop and launch the videotex service with Knight-Ridder.

My words of warning that the real evidence was mostly negative for videotex went unheeded and were perceived as a threat to the future of the organization that had been created to perform the trial and to promote the service. I even remember that an AT&T employee working on the videotex project wept when I stated the evidence for my negativism toward videotex.

Indeed, promotion has its place in the conduct of a trial of some new service. But evaluation must be independent. Hiring outside evaluators is not a solution because most consultants will avoid negative findings that might jeopardize their future relationship with the sponsor. One solution is to use a group of internal evaluators who report through an independent organizational structure to the highest levels of management. But if top management becomes falsely convinced about the success of some new product or service, few evaluators will want to suffer the fatal fate of the messenger bearing the bad news.

LESSON FROM CERRITOS: DON'T CRITICIZE

In a front-page article in the August 31, 1993 edition of the *Los Angeles Times,* John Lippman quoted me as summarizing the GTE Cerritos experiment as: "It's bombed." The reaction of GTE California's headquarters was quite strong. A GTE employee visited me and attempted to convince me that my views were wrong. I was shown a strongly worded multiple-page letter written by a GTE executive to the *Los Angeles Times* complaining about their article. A GTE employee told me that Mr. Lippman frequently got his facts wrong and was simply biased against GTE. A reporter informed me that GTE was claiming that I had been misquoted by the *LA Times*. A rumor reached me that GTE was considering writing a letter to the president of my university complaining about me.

Perhaps all these events were just circumstance, but it seemed to me that GTE California employees were over-reacting in a coordinated and very protective fashion as they conducted an apparent campaign to intimidate and discredit me.

What this demonstrates, and confirms again, is how a trial achieves a life of its own. The employees working on the trial will unite to destroy any criticism of the trial. In the case of AT&T's picturephone and of the videotex trial with Knight-Ridder, negative results were stated in a neutral or even positive way to keep the project alive. I do not know how GTE employees explained the results of the Cerritos project to their management.

PUBLICITY

There is usually much publicity when a new trial is announced and also during its conduct. However, there is usually no publicity at the end of the trial, particularly when the results are negative. Rarely are there any published results of the trial.

While I was at AT&T working on the videotex trial with Knight-Ridder, I was able to publish two papers about the trial, but in both cases I was not allowed to disclose openly the actual results and data. I can well understand the reluctance to publish proprietary results and data that would help one's competition. But knowledge is generated by a trial, and that knowledge, if published, can help others avoid pitfalls and failure. Years later, the actual data and results have little or no proprietary value, but even then most firms are reluctant to publish the real results, or allow outsiders any access to the data.

Publicity about a trial pre-positions the participants to be positive and thus can destroy the validity of the trial in terms of obtaining any realistic market data. Much earlier in this book, I mentioned the Hawthorne effect, in which simply changing the environment or showing attention to people, positions the participants positively. Clearly, what this means is that if a trial is intended to obtain meaningful data, there should be no publicity or public announcements. However, I suppose that many "trials" are really little more then thinly disguised public relations stunts, such as the GTE Cerritos "trial" and many of the Baby Bell "trials" of new video services.

The original plan for the videotex trial conducted by AT&T and Knight Ridder was to have as many information providers as possible to create a critical mass of information services. Since AT&T was a publisher of phone books through its local Bell companies, the people at AT&T who were planning the trial wanted to include the AT&T directory organization in the Knight-Ridder trial. The problem internally was that the directory organization did not want to be included because they believed they already knew that consumers wanted access to electronic phone books and also because they saw the possibilities for competition with Knight Ridder over advertisers.

The directory folks at AT&T thus decided to go it alone and announced their trial for Austin, Texas. The pre-trial publicity was intended to "place a stake in the ground" that the provision of electronic directories was a legitimate AT&T business. I remember that a colleague and I were concerned that the publicity would only serve to excite AT&T's enemies. We were right. Ultimately, as a response to the planned trial in Texas, the American Newspaper Publishers Association (ANPA) obtained an injunction against the trial and then lobbied successfully for including words in the 1982 Modification of Final Judgment to forbid AT&T and the Bell phone companies from providing information over their networks. This story points out how pre-trial publicity can be damaging in ways other than just skewing results by predisposing consumers positively.

FOCUS GROUPS

I recall a focus group interview that I attended in New york City while I was at AT&T investigating the market for video teleconferencing. We wanted to determine why a number of business people who had a free demonstration of our intercity, public room, video teleconferencing system had not used it since then.

The dozen or so participants in the focus group thought that the system worked very well and was well designed. One participant had a meeting in Chicago and had flown there for the meeting, but the plane couldn't land because of bad weather and hence he returned to New York. He flew to Chicago the next day. When asked by the moderator why he didn't use the teleconferencing system, the participant looked quite puzzled and could not express a rationale reason.

I remember another experience with focus groups, this time while I was working at the AT&T Consumer Products organization. We became quite concerned that we had no new products and that the new-product funnel was empty. We therefore wanted to talk with groups of consumers to understand what they did in a typical day along with any problems they encountered. We hoped that this would help our inventors to determine where telecommunication solutions would help. This type of unstructured study was clearly well suited to the focus group methodology.

The participants discussed their daily activities from rising from bed in the morning to falling asleep in the evening. One discovery for me was that most housewives had the TV on all day but were not watching it. They used it as a clock to know when to expect the kids home from school and as a friendly companion to listen to during the day. I thought that radio had missed the opportunity of its old role as a daily companion, but this was not a business opportunity for AT&T.

Another discovery was that a number of teenagers, and a few parents, took the stereo loudspeakers in the bathroom to listen to music while taking a shower. The opportunity here was for a stereo for the shower. A year or so later, one of the Japanese manufacturers produced such a product.

Focus groups indeed have their place in the repertoire of tools to gauge consumer needs and reactions. However, the validity of the results depend much on the skills of the moderator. Given the highly subjective nature in which the interviews are conducted and in which the participants are chosen, results are not statistically significant and should not be to projected to a wider population. Just because 50% of the participants in a focus group like some new product or service, it does not mean that 50% of the general population will hold similar views. The opportunity for an unscrupulous moderator to position the participants and hence slant the results to what the client wants to hear is a clear danger that should make you always suspicious of overly positive comments from a focus group.

At AT&T, I initiated a project to develop an inexpensive home terminal that could be used for both voice telephony and also to compose and receive digi-text messages. Our engineers designed and constructed a few prototypes of this e-mail terminal within less than a year. We then showed the terminal to a number of consumers in focus group interviews. The participants liked the terminal and associated it with a high-tech consumer-electronics firm. However, they were very uncertain how they would actually use it at home or in small businesses. Given this negative and uncertain response, I was forced to recommend that the project be terminated. Furthermore, the estimated costs of manufacture were nearly $1000, which was far above my objective of a retail price of about $100. I can still remember the tears in the eyes of the engineer responsible for the design of the prototype terminals when I told him that the project would be terminated.

What this shows is that market research can be conducted that generates the correct conclusion. However, an open-minded attitude is required and one must be willing to terminate a personally favored project. It is always important, however, to learn from such experiences, even when negative. In the case of my digi-text terminal, we learned that much education about its use would be needed for most consumers and also that it would need to sell for about $100. Given today's widespread use of personal computers in the work place and at school, many people now know about text massaging. The price objective of $100 is probably still valid though. I continue to be a firm believer in a home communication terminal that combines voice telephony with digi-text. The challenge is manufacturing and selling such a terminal for about $100.

"GEORGE MIGHT WANT IT"

The determination of consumer reactions and responses to a new product or service can be essential in avoiding big mistakes and in seizing new opportunities. Yet, when the service is truly novel, the conduct of a trial would seem to be appropriate. But, as we have seen in this chapter, independent evaluation and constant vigilance against promotion is needed.

I recall many cases when the participants in a trial or demonstration of some new product or service would say that "George might want it." When someone says this they are really saying that they themselves have no real use for the product or service, although they might actually think the product or service is a great idea. These kinds of responses are indicative of potential market failures and were the kind of responses obtained to AT&T's early picturephone product and videoconferencing service.

POTHOLES ALONG THE SUPERHIGHWAY

We have come a long way down the superhighway, have fallen into many potholes, and have come upon many collapsed overpasses. Given all the promises and wonders of the superhighway, you might by now have become somewhat depressed about its real prospects. I would not want to end this book on such a negative and depressing image. So, in the next chapter, I will share a secret with you, namely, the superhighway is already here and actually has been for many decades.

Chapter 12

A SECRET: THE SUPERHIGHWAY IS ALREADY HERE!

It is impossible to compete with the fantasy of the superhighway in terms of its technological Utopia. However, let me share a secret with you: much of the superhighway is already here, and it has been developing and evolving over the past 100 years! It is today's network of networks comprised of the public switched telephone network, the coaxial cable of CATV, communication satellites, and packet-switched networks for data communication. We tend to ignore, forget, and take much for granted that which is most visible. We tend to forget the telephone network, facsimile, e-mail, and cellular phones.

Ralph Lee Smith introduced the term "wired nation" into our vocabulary in 1970 when he envisioned and proposed a nation connected by a broadband highway of coaxial cable promising a host of telecommunication-based services. In 1987, he admitted that "the wired nation already does exist." Amen. In this chapter, we shall see how today's superhighway also already does exist.

THE INTELLIGENT NETWORK

The public telephone network goes everywhere and is essential to business in today's information economy. The telephone network is so ubiquitous and easy to use that we take it very much for granted. Yet usage continues to grow. The Baby Bells report increases in the number of access lines of over 3 percent per year and increases of about 10 percent in long-distance usage per year.

The telephone network is a switched network. The switching machines of this network are computer controlled, offering much intelligence, flexibility, and functionality to the user. Telephone networks of the distant past were operated by humans. These humans, though slow and costly, were intelligent and offered much functionality to the user. Callers could be announced by name, messages taken automatically, and calls forwarded. Human operators were replaced ultimately with electromechanical switching machines, which though offering great improvements in productivity eliminated any intelligence and functionality.

With today's use of computer-controlled switching machines, intelligence along with functionality has returned to the telephone network, but on a vast scale for everyone. Some examples are call forwarding, call waiting, and caller-ID. These "smart" services will become even fancier and more useful in the future. For example, newer caller-ID services transmit the name of the calling party rather than only the calling party's telephone number. New telephones will have a visual display so that the user can see the caller's name, and the same display could be used for text messaging and access to other test-based information services. Voice mail already allows you to receive messages even when your phone is in use.

FACSIMILE

The invention of facsimile in the 1840s actually precedes the invention of the telephone in 1876. Whereas the telephone spread throughout society, the diffusion of facsimile was quite slow. In 1977, there were only about 130,000 fax machines in use in the United States. While not a clear-cut failure, facsimile certainly was not an overwhelming mass-market success.

In the last ten years, fax has achieved much larger acceptance and market penetration. This is for a number of reasons. In the distant past, a fax machine was compatible only with the other machines made by the same manufacturer. This changed with the development of standards so that any fax machine could communicate with any other fax machine.

In the past, many minutes were needed to transmit a single page. With the new standards and improvements in technology, a page could be sent in a minute or less. Past fax machines were very costly; new fax machines are affordable by many more people, although still not affordable enough for every home. Fax machines of the past used smelly special paper that quickly discolored and faded; today's fax machines use either normal paper or other more acceptable substitutes to the smelly paper of the past.

New uses for fax seem to be invented each day. Nearly every fast food store has its fax machine so orders can be faxed there and then the food either picked up or delivered. Newsletters are routinely sent by fax. *The New York Times* offers a fax service for receiving tax forms for $5 each, but you still need to fill them out yourself. Facsimile sends a copy, or "facsimile," of the material on a piece of paper. A more efficient way to send text is as characters using 8 bits to specify each character which is the way information is stored in your word processor program. Such transmission of text is called e-mail.

E-MAIL

The telegraph preceded the telephone, and investigations of ways to place more telegraph signals on a single wire ultimately helped lead to the invention of the telephone. The telegram is virtually gone as a means of communication in much of today's industrialized world. The telegram required a knowledge of morse code and could not compete with the use of natural speech as a means of electronic communication over distance.

Today's electronic mail (or e-mail for short) is really an extension of the telegram of the past, made more ubiquitous through the use of personal computers and the telephone network to reach nearly anyone with a computer and a modem. E-mail therefore is not revolutionary but is an evolution of the telegraphy of the past to a much wider audience of users.

PRODUCTIVITY

The productivity and operating efficiency of local telephone companies continue to grow because of continued advances in technology and growth in access lines. In 1988, Pacific Telesis employed 49 people for every 10,000 access lines. In 1994, this measure of productivity had improved to 32 employees per 10,000 access lines. Pacific estimates that productivity for 1997 will be 24 employees per 10,000 access lines. When speaking to telephone company executives, I jokingly ask who will be the last employee in 2010 if this trend continues. This ever increasing productivity is the result of advances in technology.

TECHNOLOGY EVOLUTION

A decade ago, I always enjoyed taking my students on a field trip to a telephone company central office. There was no better way to understand switching than to see all the mechanical switches twisting and turning as they connected one party to another. But the last trip to see electromechanical switching was to a GTE office in Santa Monica, California many years ago, where my students and I had the wonderful opportunity to see one of the last working step-by-step Strowger switching machines. I pleaded with GTE to keep the machine and central office as a museum of telecommunication.

Today, electromechanical switching is totally gone in the United States. All switching is computer-controlled with electronic switching utilizing either space-division switching or time-division switching. Space-division switching is used in the AT&T No.1 ESS® and No.1A ESS® switching machines. Space-division switching machines are becoming the way of the past as they are extensively being replaced with time-division digital switching machines, such as the Northern Telecom DMS® and AT&T 5ESS® switching machines.

This evolution in switching has been very exciting and is an example of technological progress in the provision of telephone service. However, the twisted pairs of copper wire are a testament to technological stagnation. Although copper does not wear out from carrying electrical signals and has much bandwidth over short distances, twisted pair is costly to maintain and reconfigure for new customers. The local loop will probably be the next frontier of technological advance over the coming years.

TELEPHONE USAGE

It is futuristic to speak of some form of a new telecommunication infrastructure at the local level for telephone service. The fact of the matter is that the local infrastructure is already in place and is underutilized. The average telephone line is used only about 50 minutes per day: 37 minutes for local calls and 13 minutes for toll calls. Actually, the total level of usage has not changed that much and was 47 minutes per day in 1980, with 39 minutes of local calls and 8 minutes of toll calls.

The challenge for the telephone industry is to invent new ways to increase the usage of this valuable information-age communication system, although given all the things each of us does in a day, there simply might be little more time left for talking on the phone. The use of telephone lines for remote meter reading is one example of increased use of the existing telephone line.

REMOTE METER READING

With all the attention being given to broadband, it is exciting to discover a company that is using simple narrow-band radio in a cellular configuration to collect utility meter readings. CellNet Data Systems, Inc. has developed the service and sells it to utility companies. A small low-power digital radio is installed in the meter, and raw data is collected at many microcell receivers covering the entire service area. The data is then forwarded to a cell master receiver and then to the utility company. CellNet operates as a total end-to-end system provider, from providing the radio equipment in the meters to operating the entire cellular data network. The key to CellNet's success is providing a low cost solution to the meter-reading problem. The system can be used to read electric, gas, and water meters and can also provide various security and emergency services.

Nearly 175 thousand customers of the Hackensack Water Company in New Jersey have their water meters read remotely through a system supplied by Bell Atlantic. The system, called HOMR for Hands Off Meter Reading, includes a meter interface unit that is connected to each customer's phone line. The unit is polled during the early morning hours by equipment at the telephone company's central office, and meter usage information for each customer is then forwarded to the water company. It is claimed that with HOMR it costs only 5 cents per month to read each meter.

VOICE MAIL

Voice mail located in the network is a very useful service since with it messages can be recorded even while your phone line is busy. However, why should I pay for others to leave messages for me. The person attempting to reach me should pay to leave a voice-mail message. AT&T already offers such a service to its long-distance customers when a line is busy. You leave your message and the network attempts to reach the busy party once every half hour for six hours. You are charged $1.75 only if the called party finally receives the message.

Voice mail can be a great way for parents to call school to determine what assignments are due the next day. Used this way, voice mail substitutes for written notes from teachers to parents.

WIRELESS COMMUNICATION

Cellular telephone service has been a great success. The average length of a cellular call is about 2.4 minutes, and the average monthly bill is about $61. Cellular is only one example of wireless communication; the cordless telephone used within the home is another.

Some people believe the wireless market has much room for future growth from new technologies, one of them being PCS.

PCS

Cordless telephones are used in the home, but do not work if you go a short distance outside the home. Cellular telephones with their roaming feature can be used outside the home and all over the city and country, but they are still somewhat costly. A possible new opportunity is a form of wireless communication that fits in between cordless phones and conventional cellular.

Personal communication service (PCS) is a form of wireless that works outside the home or office, but does not have all the sophisticated features of cellular. PCS proposes to use low-power radio transmission with very limited capability for roaming and thus would be a lower-cost approach to wireless communication than conventional cellular. PCS would be a wireless service positioned between cordless and cellular. However, conventional cellular can be made lower in power and the size of the cells can be shrunk. Perhaps in this way conventional cellular could extend itself to the boundary of cordless thereby eliminating the need for a new PCS system.

The term PCS is sometimes used to have a broader meaning than only wireless. In this broader context, PCS means the ability of the network to reach a person at any time, anywhere. Intelligence in the network will know where you are at all times and reach you accordingly by telephone, wireless, pager, or whatever.

The price of cellular phones and service continues to decrease, and I am uncertain whether a new technology that looks so similar would have much of a chance in the marketplace. Cellular and cordless are terrestrially based wireless communication services.

Cellular is based on the concept of frequency reuse within a large metropolitan area, along with frequency switching as a user moves from one cell to another. An alternative approach is to place many communication satellites in orbit and to switch users from satellite to satellite. The aerospace and defense industries have been strong advocates of communication satellites and not surprisingly are promoting these extraterrestrial forms of mobile communication.

COMMUNICATION SATELLITES

The basic idea of using an artificial earth-orbiting satellite for communication was first proposed by the renowned author Arthur C. Clarke in 1945. It was not until 1962 that the first communication satellite, *Telstar*, was launched. Today, the earth's equator is ringed by communication satellites 22,300 miles above the surface of the earth. This specific orbit is such that the satellite takes exactly 24 hours for each circular orbit, thereby matching perfectly the rotation of the earth below and appearing stationary with respect to the earth's surface, a so-called geosynchronous orbit.

One problem with geosynchronous orbits is that it takes nearly one quarter of a second for a signal to travel from the earth to the satellite and then back to earth. Thus, nearly one half of a second is needed during a telephone conversation before you hear the other person respond, which is very disturbing. An early proposed solution was the use of satellites at a much lower orbit, but such low orbit satellites were not geosynchronous and would pass quickly overhead, thereby requiring a large number of satellites.

The widespread availability of optical fiber under the oceans has displaced the use of geosynchronous satellites for two-way telecommunication. However, geosynchronous satellites are still hard to beat for broadcasting one-way signals to large areas of the earth's surface. Hence, geosynchronous communication satellites are used routinely to send TV signals from network headquarters to affiliate stations and from CATV program originators to CATV operators. Geosynchronous satellites have much bandwidth and thus are also suitable for transmitting data, particularly in one direction. The *Wall Street Journal* was the first publication to be transmitted over a communication satellite so that it could be printed at a number of locations around the United States. Such satellite distribution and remote printing is today used by such other publications as *USA Today* and the *Financial Times*. But interest continues in the use of communication satellites for two-way communication of speech and data.

THE EXTRATERRESTRIALS ARE COMING: LEOS, GEOS, MEOS

As explained in the last section, communication satellites at lower orbits orbit the earth at faster rates and will not be geosynchronous, thereby requiring a series of orbiting satellites so that at least one is always above the specific area requiring communication service. Low earth-orbiting satellites are called LEOS, geosynchronous earth-orbiting satellites GEOS, medium earth-orbiting satellites MEOS, and elliptical earth-orbiting satellites EEOS. The extraterrestrial GEOS, LEOS, MEOS, and EEOS are coming, according to their proponents and hopeful investors, as a new way to interconnect the world with two-way communication. But are such hopes realistic or are they fantasy?

One of the earliest proposed LEOS was Motorola's Iridium®, employing 66 operational satellites in six polar orbits about 500 miles above the earth's surface. Iridium is projected to cost $3.4 billion, with operational service scheduled for 1997, and 1.8 million subscribers projected for 2001. The handheld telephones, providing voice telephony at a reduced bit rate of 4800 bps, would cost $1,000 in large quantities, and service would cost $3 per minute. Telephone calls from the ground could be relayed from satellite to satellite to reach the final destination. By mid 1994, only $0.8 billion had been raised, and Iridium has attracted competing systems before it itself even has become close to commercial service. Globalstar, owned by Loral, Qualcomm, and others, proposes 48 satellites in orbits about 750 miles above the earth at a cost of $1.6 billion. Odyssey, owned by TRW, proposes 12 MEOS at a cost of $2.5 billion. But the system that has attracted the most recent attention and big-name investors is Teledesic.

Not to be outdone by Iridium, the Teledesic system proposes to launch 840 satellites in 21 polar orbits 440 miles above the earth's surface at a cost of $9 billion. Each refrigerator-sized satellite will use sophisticated ATM packet switching and will offer variable bandwidth to meet different user needs. The industrialists Bill Gates and Craig McCaw are frequently mentioned as two prominent investors in Teledesic.

I was once asked to give my private views on the Teledesic system and concluded that there were serious uncertainties and astronomical funding requirements and that the system was not practical for the foreseeable future. My critical comments about the Teledesic system are applicable, though, to most similar LEO systems.

The following analysis of LEOS draws upon the framework developed in this book and supports my conclusion that LEOS are inherently unsustainable.

• TECHNOLOGY. We first ask whether the technology is available at hand today for LEO telecommunication. It is not, and there are still many unresolved uncertainties. Each satellite might need to handle tens of thousands of simultaneous users, and this degree of switching is comparable to packing ten central-office switching machines into each satellite. The high-frequency signals used to communicate to the satellite are highly directional and would be blocked by buildings and absorbed by rain. The microwave safety issue is also not well understood. At low altitude orbits, air friction is encountered and each satellite is projected to have a life of about seven years before falling back to earth. I guess we will all need to wear hard-hats as protection! But, keeping hundreds of satellites in precise orbits, switching signals from one satellite to another as each rapidly passes over the horizon, and keeping track of thousands of simultaneous users for each satellite are serious technical challenges, perhaps resolved only on paper, but not in yet in orbit. LEOS do minimize greatly the round-trip delay encountered with GEOS, and LEOS require much less power since they are so close to the earth. However, these advantages are few and are outweighed by serious technical problems.

• FINANCE. About 70 percent of the time, each low earth-orbit satellite will be over oceans and thus not of much use. The high cost of each satellite is thus not justified by its low average utilization. For the system to be operational, all the satellites must be in orbit, and there is no way to increase the system gradually as traffic develops, like terrestrial cellular. The financials proposed by the promoters of these systems are all overly optimistic and are far from the realism of the true costs of the technology, launch costs, and disasters. The yearly profits needed to repay the $9 billion investment required for the Teledesic system in 10 years at a return of 15 percent are $1.8 billion. It is very difficult to believe that such profits could ever be generated, given the competition from existing cellular systems and also the competition from other satellite systems.

• POLICY/REGULATORY. Licenses would need to be obtained from many of the countries of the world, and some are not liberal regarding competition, particularly from United States firms.

• CONSUMER/MARKET. The terrestrial cellular systems are already installed, are growing rapidly, and are formidable competitors to LEOS. It would be inconceivable that LEOS could compete in large cities with existing cellular. Thus the only real market for LEOS would be in areas of the planet not served by conventional telephone or cellular service. Such areas are few and most certainly are poor countries whose citizens could never afford such a luxurious service. This leads me to wonder whether the only markets for these systems are made up of tourists on safari in equatorial Africa who are about to be charged by an enraged bull elephant and need to call international 9-1-1 for help, or shipwrecked survivors afloat in a raft in the middle of the Atlantic.

Although the preceding analysis is negative regarding LEOS, communication satellites do have their place, as was discussed earlier. They are a great way to distribute network television programs around the earth and domestically. They are also appropriate for reaching otherwise inaccessible portions of the planet. For these kinds of applications, GEOS have much to offer as they are few and less costly since the satellite can be used nearly all the time.

American Mobile Satellite Corporation, partially owned by Hughes Communications, will offer voice and fax using three GEOS at a development cost of $560 million. The first of the three satellites was launched in April 1995. Portable mobile units that need to be aimed at the satellites will be used. Simple digital messages are proposed to be carried by Orbital Sciences' OrbComm system using 26 satellites.

Satellites are also a good way to obtain precise positioning information. The Global Positioning System (GPS) is already in place with 24 satellites in medium-earth-orbit, broadcasting codes to enable users to position themselves accurately. Boaters and civilian aircraft are some of the users of GPS. Some people believe positioning receivers will soon be in automobiles to guide us along the auto highways of the world. Hand-held receivers, priced at about $200, are already available to help hikers find their way back to camp.

TELEVISION

Television is perhaps one of the most exciting forms of entertainment today. TV has been characterized as being nothing more than radio with pictures. At one time, it was predicted that TV would replace movie theaters. As an entertainment medium, TV is probably taken far too seriously by the academic community which studies television to understand why people watch it and also to determine its impact on society. We all know TV is watched because it is a passive entertainment medium that is instantly available and that requires no effort on our part. Yet so pervasive has television become that very few remember the days before TV.

The key to electronic television was the use of the cathode ray tube, first proposed by Prof. Boris Rosing in St. Petersburg, Russia in 1907. His student, Vladimir K. Zworykin, came to the United States in 1919 to work at the Westinghouse Company in Pittsburgh. From 1923 to 1925, he invented and demonstrated the technology that would become the basis for today's electronic television. David Sarnoff of RCA supported Zworykin's work and in 1929 Zworykin went to RCA to lead the development of today's commercial broadcast television.

In 1945, the FCC allocated spectrum space for thirteen VHF television channels, and broadcast television was on its way. Market penetration was very fast: 6% of United States homes in 1949; 49% in 1953. A color TV system (called field sequential) that was not compatible with existing monochrome television was accepted as standard by the FCC in 1950. A few years later, in 1953, it was replaced by the compatible NTSC system invented by RCA engineers.

Today, television reaches many of us over the air by VHF (very high frequency) and UHF (ultra high frequency) broadcast radio signals. Over-the-air television reaches many places but not those places shielded by hills and mountains. The solution is CATV.

CATV

The first commercial CATV system was installed around 1950 as a community antenna TV service (CATV) to serve homes behind a large mountain that shielded them from the TV signals being transmitted from a large nearby city. A master antenna atop the mountain received the TV signals which were then amplified and sent over coaxial cable to the homes. From its beginnings as a way to obtain TV in areas distant from TV stations or cities subject to propagation problems from many tall buildings, CATV became the "television of abundance" promising many channels of specialized and unique programming over coaxial cable.

In 1979, cable passed by 38% of TV households, and 20% of TV households actually subscribed to CATV. In 1977, the total revenue of the CATV industry was $1.2 billion with an average subscriber paying $8 per month. Today, the average CATV monthly bill is nearly $35, cable passes by 98% of TV household, and a little more than 60% actually subscribe. CATV has become truly a national mass-market service and is a major part of today's communication superhighway.

DIRECT BROADCAST SATELLITE TV

A friend of mine lives in a rural area of New Jersey and is not served by the coaxial cable of a CATV firm. It would be too costly to install cable to serve such a low density of potential subscribers. Instead, my friend subscribes to direct broadcast satellite TV offered by DIRECTV.

Direct broadcast satellite (DBS) television is based on a communication satellite in geosynchronous orbit above the earth's equator. The satellite receives a signal from the earth and then simply rebroadcasts that signal over a wide area of the earth's surface.

After many years of technical development by Hughes Communications, its DBS service, named DIRECTV™ and offered by DIRECTV, Inc., was finally launched in 1994. United States Satellite Broadcasting, Inc. (USSB) also offers DBS service over the Hughes satellites along with DIRECTV service. Between them, they currently offer 150 channels from three collocated satellites. Sophisticated digital compression is used to transmit each TV channel at an average bit rate of about 3 Mbps to 5 Mbps, depending on motion. The image quality that I saw was excellent, although I am told that there can be some blurring of fast motion in rapid-action sports events.

A small 18-inch diameter antenna along with appropriate electronics, priced at about $700, is needed to receive and decode the signal. A monthly fee must also paid for the service in addition to the cost of the antenna and receiving decoder. Movies are shown on more than one channel, offset in time, for about $3 per viewing. DIRECTV claims to need only 3 million subscribers to break even financially and hopes to have 10 million subscribers by the year 2,000. It claims to have had over 400 thousand at the first quarter of 1995.

DIRECTV and USSB have two new competitors. PrimeStar Partners provides about 95 channels on its single-satellite service, utilizing a somewhat larger receiving dish than the Hughes system. PrimeStar packages its service to include rental of the receiving equipment in a monthly fee so that its customers can avoid the up-front purchase of the costly receiving equipment. The PrimeStar satellite was launched in 1995. PrimeStar claimed to have one-million subscribers and 45 percent of the DBS market in the United States at the end of 1995. PrimeStar plans a $150 million marketing campaign for 1996 and plans to launch new satellites that will allow the use of the smaller 18-inch receiving dish.

EchoStar Communications Corporation started offering DBS service on its single satellite in March 1996. EchoStar offers its customers the option to buy, finance, or lease the 18-inch dish and receiving equipment. EchoStar plans to launch two more satellites by the Fall of 1997. The EchoStar system is called DISHNETWORK^{SM} DBS service.

DIRECTV, USSB, PrimeStar, EchoStar, and any other new entrants will have to share the market for DBS TV, and this will reduce the profitability of each. Whether any service achieves the market penetrations needed to survive financially remains to be seen, particularly given the tremendous threat of existing conventional CATV which can also use digital techniques to provide many more channels of programming.

One potential problem with satellite-based television systems is that local broadcasting and programming is not available. Thus, many subscribers to the service still need to subscribe to the local CATV service to obtain this local programming. One solution would be to broadcast a form of national localized programming over DBS in a manner similar to the local news contained in the national newspaper *USA Today*. But perhaps the importance of local news is overemphasized, and in today's global village, global news is far more important and interesting than boring, small-town gossip.

DBS has attracted the attention of telecommunication giants. In the fall of 1995, AT&T paid $137.5 million to purchase 2.5 percent of

DIRECTV. I fail to see any synergistic opportunities with AT&T's core businesses in this investment.

WIRELESS CABLE

Direct broadcast satellite television uses very-high frequency microwave radio to transmit the TV signals from geosynchronous orbit to homes. Terrestrial microwave radio can be used to deliver TV programs to homes. The basic idea is quite old and has been used for single-channel pay-TV—an approach called MDS , for Multipoint Distribution Service. Newer technology offers as many as eight channels and is known as MMDS, for Multichannel Multipoint Distribution Service.

On the positive side, wireless cable through MMDS does not require the costly infrastructure of coaxial cable. But on the negative side, microwave signals travel in a straight line and are easily obstructed by such items as the leaves on a tree or even ground fog. Newer MMDS technology is digital, offering even more channels of video programming. Some Baby Bells have been very active in acquiring MMDS systems for video.

The number of ways to obtain television seems limitless: over-the-air VHF and UHF, video tape, cable television, direct broadcast satellite, optical fiber, and MMDS. With competing versions within many of these distribution systems, it seems clear that not all will survive. Or is our appetite for video so large that all will survive along with many coming new alternatives?

SOME MISSED OPPORTUNITIES

Reality can not compete with hype and fantasy. This means that real opportunities are frequently overlooked and missed as attention is focused on nonsense rather than reality.

One way to extend a medium is to marry it with another modality of communication. A problem with broadcast radio when it comes to music is that you have to wait until the music is over until the announcer identifies what you have been listening to. A solution would be to use text broadcasted along with the music, so that the text could be displayed on a small LCD panel in your radio. In this way, you could know you were listening to Bernstein's performance of Mahler's *Resurrection Symphony*. The whole experience of radio could be improved if radios with programmable audio cassette recorders—similar to the programmable recording feature of VCR's—were available. In this way, you could record a favorite radio show for later listening.

The electronic transfer of funds is quite routine today. My pay check is transmitted electronically by my employer to my bank. The problem is that I do not know precisely when and in what amount this deposit has occurred. As a form of automatic confirmation, the bank's computer could telephone me, perhaps late at night while my ringer is off, and leave a speech synthesized message stating the date and amount of the deposit. I could then enter the information into my check book for balancing in the morning.

Many times when requesting information over the phone I waste time spelling my name and address. All this information could easily be transmitted as an electronic business card stored in the telephone. My name and address would be sent to the called party in a short burst of digital data during the telephone call.

The use of a multi-stacked recording head to record 8 bits at a time could be used with image compression to produce a video recorder that would cost about the same as an audio cassette machine.

TELETEXT: ANOTHER MISSED OPPORTUNITY

In 1981 while I was at AT&T working on the videotex trials, I attempted to segment the market for home information. My segmentation consisted of four segments: (1) videotex systems that used the telephone line for access to a large central database of information in color with graphics, (2) stand-alone computer systems for handling large amounts of encyclopedic data on an interactive basis, (3) teletext, and (4) textual telecommunication, or what is today called e-mail. Based upon what was learned from the marketplace, I concluded that videotex had poor prospects for market success, as the other three systems could do what videotex promised, but more easily.

As early as 1981, teletext looked like a coming success. And indeed it is throughout most of Europe, even though most people in the United States have never even heard of it. It is very unusual to discover a communication service that has achieved mass-market availability throughout Europe, but is virtually unknown in the United States. So what then is teletext, and why is it so successful?

Teletext sends a few hundred screens of information along with the normal broadcast TV signal by imbedding the digital data in the otherwise unused vertical blanking interval of the TV signal. The few hundred screens of text and simple graphics information are sent continuously in a round-robin fashion. You select the number of the page of information that you wish, and when it circulates around it is grabbed and captured for display on the TV set.

Teletext is great for timely mass-market information, like weather, traffic, and news headlines. It is free and does not require costly equipment or an elaborate rewiring of the telecommunication infrastructure.

Teletext is available throughout Europe, but is not available in the United States. Teletext can be a textual adjunct to broadcast television and can identify what show or movie is being watched. Teletext can be used to give local advertsing information as an adjunct to a national commercial. Teletext can give information about how to order something advertised in a broadcast commercial.

Teletext is an information service that does not promise all the world's information at your fingertips. Teletext offers fast access to a limited database of timely, general-interest information. It uses the home TV set in a friendly, easy-to-use fashion and is free. Why then don't we have teletext in the United States?

One reason is that the standards question became complicated when a sophisticated graphics system called NAPLPS (North American Presentation Level Protocol Standard) was proposed for teletext. NAPLPS was simply far too costly and complicated. Another reason is that someone has to make a profit for teletext to be successful.

Although I believe teletext would be profitable, the profits would certainly be small compared to the profitability of television in general. Most of the benefits would accrue to set manufacturers, although the local affiliates and networks would make additional revenue from new forms of more useful advertising. Stereo audio has no impact on broadcaster profits and although the only new profits are to set manufacturers, we do have stereo sound for TV. European governments mandated that teletext decoders be included in all TV sets, and this had a large impact on the widespread availability of teletext service. In the United States, teletext simply can not compete with all the hype over multimedia, virtual reality, high-definition TV, and cyberspace; the final result is that the teletext opportunity continues to be ignored.

ECONOMIC OPPORTUNITIES

Technology is not the only driving force in shaping the future, as we have seen. Finance and economics can also have a large effect on the future. Long-distance telephone rates have traditionally been based on mileage bands, with increasing rates for increasing distance. On the other hand, domestic postal service has, from the earliest days of the "penny post," been based on a single flat rate regardless of distance. Furthermore, post charges are based on weight, and long-distance charges are based on the length of a call. AT&T has initiated a new charging option for domestic calls in which the rates are flat regardless of distance, just like the post. Such a flat-rate charging plan is only possible because of network efficiencies and economies of scale.

False economic opportunities are also possible. We saw in the chapter on the Internet that federal subsidization and flat rates have created false economies that discourage private entrepreneurship and competition.

READINGS

CATV

Foster, S. Ronald, "CATV Systems are Evolving to Support a Wide Range of Services," *Telecommunications*, January 1994, pp. 95-96, 98.

Landler, Mark, Ronald Grover, and Kathy Rebello, "Cable's Bright Picture Fades To Gray," *Business Week*, May 30, 1994, pp. 130-132.

Wittering, Stewart, "Cable Networks and the Local Loop," *Communications International*, November 1992, pp. 50-53.

Wireless

Arnst, Catherine, "Talk About Dialing For Dollars," *Business Week*, January 9, 1995, p. 78.

Flanagan, Patrick, "PCS Spectrum Auction Sets Stage for Wireless Wars," *Telecommunications*, May 1995, pp. 17-18.

Flanagan, Patrick, "Personal Communications Services: The Long Road Ahead," *Telecommunications*, Vol. 30, No. 2, February 1996, pp. 23, 24, 26, 28.

Lewyn, Mark and Peter Coy, "Airwave Wars," *Business Week*, July 23, 1990, pp. 48-53.

Padgett, Jay E., Christoph G. Günther, and Takeshi Hattori, "Overview of Wireless Personal Communications," *IEEE Communications Magazine*, January 1995, pp. 28-41.

Taylor, Paul, "Technology for the next generation," *Financial Times*, September 5, 1994

"Mobile Telephones," *The Economist*, May 30, 1992, pp. 19-22.

Satellite PCS

Emmerson, Andrew, "You can always switch off," *Financial Times*, October 17, 1994.

Frieden, Rob, "Satellites in the Wireless Revolution: The Need for Realistic Perspectives," *Telecommunications*, June 1994, pp. 33-36.

Frieden, Rob, "Satellite-based Personal Communication Services," *Telecommunications*, December 1994, pp. 25-28.

Lewyn, Mark, "He's No Mere Satellite-Gazer," *Business Week*, April 4, 1994, p. 39.

Taylor, Paul, "New breed of satellites into orbit," *Financial Times*, September 5, 1994.

Van, Jon, "An airy quality to satellite plans," *Chicago Tribune*, March 25, 1994, Section 3, pp. 1,4.

"Infrastructure in the sky," *The Economist*, March 26, 1994, pp. 101-102.

"Beam me up, Scottie," *The Economist*, March 28, 1992, pp. 69-71.

Other Services

Bulkeley, William M., "The Videophone Era May Finally be Near, Bringing Big Changes," *The Wall Street Journal*, March 10, 1992, pp. 1, A4.

Davis, Bob, "U.S. Looks Into Global Radio Network Using Satellites," *The Wall Street Journal*, June 26, 1990.

Noll, A. Michael, "Videophone: A Flop That Won't Die," *The New York Times*, January 12, 1992, Section 3, p. 13.

Noll, A. Michael, "Anatomy of a failure: picturephone revisited," *Telecommunications Policy*, May/June 1992, pp. 307-316.

Singh, Indu B., "HDTV: Technology and Strategic Positioning," *Telematics and Informatics*, Vol. 8, Nos. 1/2, 1991, pp. 1-8.

"ISDN: Back to the future," *The Economist*, January 7, 1995, pp. 54-55.

Chapter 13

DREAMS OF THE FUTURE

We saw in the previous chapter that today's telephone, data, and video communication systems already form a communication and information superhighway that enables us to communicate nearly anywhere, at anytime, with anyone by a variety of modalities such as speech, text, graphics, and video. But what should be our dreams of the future?

TECHNOLOGY

Clearly, technology has and will continue to have a major impact on shaping the future. Technology will continue to advance, sometimes in ways that can not be currently foreseen.

Computer control will continue to have much impact on the telephone network. Optical fiber and time division multiplexing will likewise continue to impact the telephone network, leading to lower costs and increased reliability.

Fiber offers much bandwidth. There has always been a contest between bandwidth and compression. Decades ago, speech compression devices, called vocoders, were able to compress a 4,000 Hz telephone speech signal to as little as 200 Hz. However, the technology to do so was complex and costly. Advances in bandwidth always occurred more readily, and hence speech compression was rarely used in the telephone network.

The transmission of color and elaborate graphic images over the Internet makes the system more friendly to users but consumes many bits. The actual information is considerably lower in bits. It would thus make sense to send the information with as few bits as possible and then use software at the user's computer to construct the friendly interface with graphics and color.

Some basic principles of technology can not be easily violated. Signals occupy spectrum space, called bandwidth. Bandwidth costs money, even if very little money, and the bandwidth of one signal usually displaces another.

We know from the past that clever advances will occur in the invention and development of unforseen technologies. If there is anything that history teaches, it is that whatever is much in vogue today will most likely disappear tomorrow. Most major advances were not viewed as important when first invented, and many advances in technology were first commercially applied in ways unforseen by the inventors. One example is linear predictive coding (LPC), invented at Bell Labs in the 1960s. The supposed application was in speech processing to save bandwidth on telephone circuits. However, the first commercial use of LPC was by Texas Instruments in a spelling toy. Bell Labs completely missed this application, one reason being that they did not view toys as their market. Another example from Bell Labs is the pioneering use in the 1960s of computers to perform the animation of title sequences for movies and videos. Again, Bell Labs did not patent nor commercialize the technology, because it did not view the arts as its business.

Advances in technology frequently take decades to be developed commercially and to reach market success. Given the shortsighted interests of many large firms today, few of them would be willing to take the necessary risks of such long-term investment. Basic research will suffer, and the cupboard of basic knowledge will slowly empty. But shortsightedness can be short lived too, and a return to the adequate support of basic research will occur. Technology is simply too important to ignore for too long.

Technology is only one of the five factors that will shape the future. The others are finance, consumers, business, and policy, and advances in these areas can have big impacts.

CONSUMERS

Consumers have basic needs that change little, such as: the need for shelter and housing, the need for food, the need for security, the need for transportation, the need to be entertained. Although these basic needs do not change over time, how they are satisfied indeed can change dramatically. For example, the need for transportation that was satisfied by the horse is today satisfied by the automobile and the jet airplane.

The telecommunication behavior of consumers is little studied and understood. We still do not fully understand why the visual imagery of a videophone is so distasteful to most consumers. We do not understand why the intimacy of a telephone conversation can be so easily destroyed by a speakerphone. Nor do we understand why the fidelity of sound is so much more important than the fidelity of video for entertainment.

An increased understanding of consumer behavior might enable us to predict more correctly what products and services will succeed or fail. A model of telecommunication behavior might even enable us to invent new products and services. It will be interesting to see whether the research necessary to develop such understandings and models will be supported and performed. In the past, social and sociological research has been viewed with skepticism and scorn by most business people and technologists.

In the end, it is always consumers who decide the future. If they use and like new products and services, then the new will survive. However, new products and services involve changes in old habits. The new must be easy and understandable to use. The Internet achieved mass interest only when friendly interfaces had been developed.

FINANCE

Pouring big bucks into big projects usually results in big losses. This is very true in basic research. A better strategy would be to undertake many small risky projects. It is impossible to know in advance which ones will succeed—such is the nature of innovation and research.

When I was at the White House Science Advisor's office, I once suggested, tongue-in-cheek, in a memorandum to the Office of Management and Budget that the National Science Foundation cease its large grants and peer-review mechanisms and instead simply give away its then $1 billion budget as a series of small gifts of $20,000 each to the first 50,000 academics in line at the NSF headquarters building. Administrative costs would be vastly reduced, researchers who otherwise would never be funded would receive funding, risky research could be undertaken, and all the wasted time and effort of writing proposals would be avoided. The NSF budget has now grown to $3 billion, which would be enough to grant each of the estimated 200,000 academic scientists and engineers $15,000 a year. But big science with its big, million-dollar grants has become a major source of revenue for many big universities—they would not relinquish this pork-barrel.

POLICY

The short-term future will most certainly see a greater concentration of control in the hands of a few companies. Early in 1996, SBC Corporation (the old Southwestern Bell) announced its intent to acquire Pacific Telesis, and a month or so later, Bell Atlantic and NYNEX announced that they were planning to merge.

The Baby Bells will continue to merge, perhaps because they believe that bigger is better. However, I continue to believe that the provision of telephone and telecommunication service is a natural monopoly, and hence the mergers of the Baby Bells to recreate the old Bell System can be viewed simply as gravity causing the entities to come together again. The merged Baby Bells will enter intra-state long distance, perhaps acquiring AT&T or one of the other large long-distance companies.

And so the old Bell System reforms. Is this a problem? One big difference is that its recreation will include control of CATV (wired and wireless) and its video content. Also, regulation has been greatly reduced, thereby opening the door to unregulated monopoly. Indeed, in the short term all sorts of crazy things can happen, but in the end, sanity usually prevails. In this specific re-formation of the old Bell System, the Antitrust Division of the United States Department of Justice will probably ultimately prevail and will force some form of restructuring on the merged Baby Bells, perhaps requiring them to relinquish their domination of CATV and entertainment.

MULTIPLEXING IN LOCAL LOOP

A visit to a telephone company central office is most enlightening. The first thing you notice is all the empty floor space that used to be occupied by the electromechanical switching systems of the past. The new computer-controlled digital switching systems, either the blue-and-gray cabinets of AT&T or the brown-and-green cabinets of Northern Telecom, occupy considerably less space. You next notice the total absence of any people servicing these switching systems. The machines are monitored and controlled remotely from central service bureaus. Lastly, you notice a long aisle of the main frame consisting of wires that interconnect the thousands of wires of local loop pairs to the switching system. Usually, you find the only humans here moving wires around. Clearly, in this day of modern electronics, all these wires do not make sense.

The opportunity is to utilize multiplexing in the local loop so that thousands of voice circuits are brought together to the central office on a single medium, such as optical fiber. As one example of the potential impact of this new local architecture, Pacific Telesis estimates cost savings of $50 per line in providing telephone service from the use of its new integrated video/telephony network. These cost savings alone would, in Pacific's view, justify the investment in the new local architecture.

The issue here though is timing. When will the multiplexing technology really be cost effective? Today's telephone network has centralized intelligence and centralized provision of the direct current needed to power telephones. This approach is simple, robust, and survivable in time of disaster. The use of sophisticated electronics in the network will indeed someday make good sense, but not if it reduces reliability and survivability and is far too costly.

Progress sometimes is very slow. In the mid 1970s, I wrote to the chairman of the board of AT&T to suggest an intelligent interface between the local loop and the distribution wires in homes. This interface would facilitate loop testing to determine whether customer telephones or the loop itself was at fault in service problems. Since this was before the Bell breakup and at a time when AT&T was still fighting competition in the provision of telephones, I was told that my idea made no sense. Two decades later we still do not have such an interface.

As progress in technology continues to shrink in physical size, the question arises about what to do with all the unused space in most central offices. Some telephone companies are using the space for various clerical operations, although much renovation is usually needed since the space was designed originally for machines and not for the comfort of humans.

THE FUTURE

The future will certainly see a continuation of the many telecommunication services that we take mostly for granted today: the telephone, facsimile, cellular, and pagers.

Growth and interest about the Internet will certainly continue, but with far less hype. Packet switching will become ubiquitous with a number of public packet networks offering service, assuming that telecommunication does not return to a natural monopoly in the United States. E-mail will also be widespread with the availability of inexpensive dedicated e-mail terminals. These "digi-text" terminals will be integrated with telephone service and will help us control the intelligent features of the telephone network.

The price of domestic long-distance will continue to decline at a rate of at least 4% per year. The price of international calling will likewise continue to decline, and indeed the world will continue to become smaller because of affordable telecommunication.

Local data networks will be developed to increase the efficient use of the local loop. Multiplexing of telephone lines will occur in the local loop to increase the engineering efficiency of the telephone system.

After a wild ending to the twentieth century with much merger-mania and hype over media concentration, the beginning of the twenty-first century will look much like today in terms of industry structure and boundaries.

It seems easier for me to state what will not occur than to predict what will occur. I guess that this means that it really is impossible to predict the future. Much of what we take for granted today was never foretold. And much of what was foretold never occurred.

I have presented many critical perspectives in this book, but I have also presented a methodology for making reasoned opinions about new products and services. A negative view can be very valuable. Avoiding mistakes and big financial losses can be viewed as profit. Proving that repeats of mistakes of the past do not make good sense is not that challenging. The burden of proof should be on those who propose "new" ideas.

Chapter 14

A FINAL PERSPECTIVE

We have finally come to the end of our long journey along the superhighway. It has not been as super as many have promised. The many potholes and washed out bridges have jolted our sense of reality. The promised gold at the end of the superhighway rainbow has been as illusionary as always.

We saw that history again repeats itself. We saw that the five necessary links to create a viable venture—technology, finance, consumer, business, and policy—either were very weak or were broken. We saw that the superhighway is a multimedia, nonsensical dream of silly hopes. But why then is it given so much attention by so many large and supposedly sane companies and their top executives? As a final perspective, I will present several possible answers to this question.

Some futurists, who are paid much money to advise communication companies, state that the future and real profits are in the provision of entertainment content. Telephone companies might actually be guided by this questionable advice and therefore are simply lusting after Hollywood and CATV.

We saw in an earlier chapter that the financial implications of the Bell Atlantic planned acquisition of TCI made no sense for Bell Atlantic. TCI, however, understood well that through the sale of the company it could unburden itself of much weighty debt. It would therefore seem that poor business judgement was clouding Bell Atlantic's judgment in its blind lust for TCI. US WEST has acquired a 25% interest in Time Warner Entertainment—another company with questionable financial performance. In this case, US WEST's motivation appeared to be guided by a strategy to link with content, in effect, an overfascination with "Hollywood"—a form of Hollywood-mania which in the past has trapped many people into losing fortunes.

When I talk to telephone company executives, I discover that they know little of the history of the past decades of failed attempts to combine television and telephony on a single medium. I find much ignorance of the failed picturephone, of technology in general, and of the challenges of measuring consumer reactions. They seem guided by an almost religious fervor of blind faith and hope. This faith is supported by their own self-generated hype and overpromotion of a technological Utopia, similar to the one presented at the very beginning of this book.

Thus we have one possible explanation for all the hype of the superhighway in terms of blind faith and false hopes. Yet we question whether the giants of the communication industry could really be this naive and ignorant. There must be other and better reasons to explain the superhype of the superhighway. The chief executives of these firms must be wiser than to fall for their own promotion and hype. One explanation is too much money seeking too few, really new ideas.

The telephone companies—the Baby Bells and GTE—make tremendous profits, with two-digit after-tax profit margins. The pockets of these companies are stuffed with cash. Rather than return these profits to their customers through lower rates, the telephone companies are seeking new ventures and investments. The local telephone companies are concerned that their local monopoly core business is in risk of losing profits. Hence, as a hedge against such competition they clamor to be allowed into manufacturing, long distance, and entertainment. Meanwhile, software and media companies are convinced that entertainment, television, and computers are converging, creating opportunities for great wealth. Hence new alliances and mergers are formed to pursue these new supposed opportunities. Thus, good old-fashioned business greed can explain much of the hype of the superhighway. But isn't it the mission of business to promote itself to maximize profits?

There is a fear and near panic that something really big is about to happen and that one will be left behind if something is not done immediately. The telephone companies fear the CATV companies who in return fear the telephone companies. These fears are also self-fulfilling as each company attempts to outmaneuver the other through strategic partnerships and intense lobbying at the local, state, and federal levels. So, fear of being left behind, perhaps in a cloud of semiconductor dust, is a strong factor in today's media mania. There is also a lack of imagination and a desire to follow the pack, both of which follow from a fear of risk.

Politics is also a factor and is being played at the highest levels. The plan is to dangle a nebulous vision of a technological Utopia before the politicians and invent and embrace "new" concepts such as a National Information Infrastructure or communication superhighway and then ask for less regulation so that the Baby Bells can build the new infrastructure. Decades ago, the CATV companies pursued a similar strategy by promising all sorts of nebulous, interactive, two-way cable services along with community access. They would promise anything to obtain franchises but then never delivered on the promises. We can only hope that the politicians and regulators will ultimately see through today's reincarnation of false promises, or that the plan will self-destruct because of its own greedy complexity and chaos.

Although I have attempted to convince you of otherwise, perhaps you still are convinced that the communication superhighway is coming and will be the dawn of a new, revolutionary era and society. I hope that our journey has educated, informed, and convinced you that the superhighway is mostly superhype. Although I have attempted to give some possible explanations, in the end, I have no real single explanation of all the hype of the superhighway. Blind faith, ignorance, greed, fear, false hopes, naivete, poor business judgment, amateur politics—take your own pick from these. But whatever you do, avoid the on-ramp to the superhighway and keep to the much safer surface streets!

Appendix A

TECHNOLOGY PRIMER

Technology is a major factor in determining and shaping the future. Innovations in electronics have made possible such essential parts of our lives as television, radio, telephone service, recorded music, and the word processor. A basic literacy of some of the core concepts of modern electronics is essential if one is to understand the role of technology in shaping the future. This appendix explores some of these key concepts and thus is a precursor to discussing the specific role of technology and the various technological issues involved in the superhighway discussed in Chapter 4.

In this appendix, we shall see that certain fundamental technological principles involving bandwidth are essential in shaping the future. We will learn about the process of digital encoding that has changed forever the nature of audio and telecommunication and what it means for the future. And lastly, we will discuss optical fiber as a transmission medium and its use in networks. The three technological topics of bandwidth, digital, and optical fiber are mentioned most frequently as essential for the superhighway and it is important that we learn what they are really all about. We start with signals and how to characterize and describe them.

SIGNALS

The whole world of modern electronic communication involves the conveyance, storage, and processing of signals. A signal varies with time, as shown in Fig. A1. The Morse code of the telegraph involves electrical signals that are created at the telegraph key as interruptions in the flow of current and then cause the audible clickety-clack at the receiver.

Human speech creates an acoustic signal that is transduced into an electrical signal by the microphone within the telephone handset. This electrical signal is then conveyed over great distances by various transmission media, such as copper wire, coaxial cable, radio, and optical fiber. At the distant end, the signal is transduced back into sound by the small loudspeaker in the telephone handset.

A sound wave can be saved and imprinted on a phonograph record. The squiggles in the grove along the phonograph record are microscopic replica of the actual sound wave. This replica is an analogy of the original waveform—hence the term analog describes signals that look like the original waveform.

Sound waves are variations in sound pressure. These variations when shown or plotted versus time have different shapes, called waveshapes or waveforms. Engineers look at these waveforms using a device called an oscilloscope, which is a monitor for examining waveforms. A waveform usually has a maximum peak in the upward or positive direction and a maximum peak in the downward or negative direction. Many waveforms are symmetric with identical positive and negative shapes. Other waveforms, like the hiss encountered between radio stations, are purely random noise with no discernible pattern. The excursions of the waveform are called the amplitude of the waveform.

Some waveforms have a fundamental shape that simply keeps repeating itself over time—such wavefroms are called periodic—as

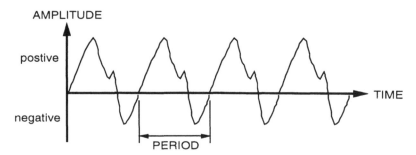

Fig. A1. All signals can be represented in terms of their variation with respect to time. The shape of this variation is called the waveform or waveshape of the signal. The excursion of the waveform is called amplitude and can be in the positive and negative directions. Some waveforms—called periodic—have a fundamental shape that keeps repeating in time.

shown in Fig. A1. The square wave—so called because the fundamental shape looks like a square—is a periodic waveform. The rate at which a waveform repeats itself is called its frequency, measured in the number of repeats, or cycles, per second. A more technical term for the measure of frequency is Hertz, abbreviated Hz. One cycle per second is the same as one Hertz. Hertz is frequently measured in units of thousands, millions, and billions: 1,000 Hertz is the same as 1 kilo-Hertz, abbreviated as 1 kHz; 1,000,000 Hertz is 1 mega-Hertz, or 1 MHz; and 1,000,000,000 Hertz is 1 giga-Hertz, or 1 GHz.

SPECTRUM

Looking at signals in terms of their variation and shape over time is only one way they can be characterized, and an equally powerful way is to examine their frequency components. Signals and their waveforms are composed of different frequencies and occupy space in the frequency spectrum. The spectrum of a signal shows the amounts of the various frequencies that comprise the signal in the same way that the spectrum of light shows the different light frequencies corresponding to different colors, as shown in Fig. A2. An amplifier or communication channel does not pass all frequencies equally and thus has a frequency response that shows the relative amounts of different frequencies passed.

Signals are composed of a finite range of frequencies, and communication channels and devices pass a finite range of frequencies. The width of the range of frequencies in a signal or passed by a

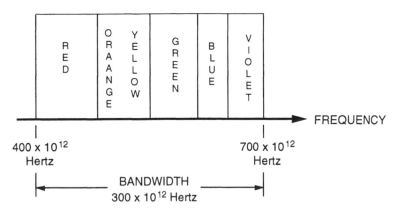

Fig. A2. The spectrum of visible light extends across a range of frequencies, or bandwidth, of 300 x 10^{12} Hertz.

communication channel is called the bandwidth of the signal or channel. The bandwidths of some signals and communication channels are shown in the table below.

SIGNAL/CHANNEL	BANDWIDTH
telephone speech signal	4 kHz
AM radio music	5 kHz
AM radio station	10 kHz
hi-fi amplifier	20 kHz
FM radio station	200 kHz
TV video signal	4,200 kHz
TV station	6,000 kHz

We see from this table that television takes considerably more bandwidth than other communication channels. Although the bandwidth of a TV channel is 6,000 kHz, or 6 MHz, the bandwidth of an actual TV video signal is 4.2 MHz—about 1000 times the bandwidth of a telephone signal. In general, the bandwidth of a communication channel must be as least as large as the bandwidth of the signals it will carry, or else information will be lost.

A hi-fi amplifier passes frequencies from 10 Hz to 20 kHz. Its bandwidth is 20 kHz. However, few humans hear sound frequencies much higher than about 13 kHz. The additional frequencies passed by the hi-fi amplifier are like an acoustic cushion just to be certain that no distortion of the sound occurs before we hear it. The band of frequencies assigned to an individual FM radio station is 200kHz wide, and all FM radio stations are within the FM radio band, as shown in Fig. A3.

MODULATION

Signals can be shifted and moved about in the frequency spectrum through a process called frequency shifting. Frequency shifting is accomplished through either amplitude modulation or frequency modulation. A thorough description of the processes of amplitude modulation and frequency modulation are beyond the scope of this book, but some of the basics of the processes are as follows.

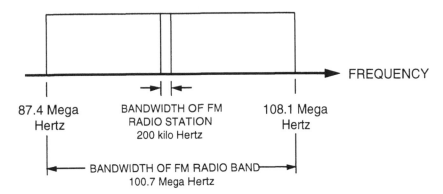

Fig. A3. The portion of the radio electromagnetic spectrum allocated for FM broadcast radio is from 87.4 MHz to 108.1 MHz—a bandwidth of 100.7 MHz. Each FM radio station occupies a band of 200 kHz, although the transmitted speech or music has a bandwidth of only 15 kHz. The additional spectrum space of each FM radio station is used for stereo information and noise immunity.

A carrier wave is used, something like a catalyst, to carry the signal from one range of the spectrum to another. This carrier wave is a pure tone of a single frequency. With amplitude modulation, the peaks of the carrier are made to vary in exact synchrony with the modulating signal. With frequency modulation, the frequency of the carrier increases and decreases slightly in exact synchrony with the modulating signal. With amplitude modulation, the bandwidth of the modulated carrier is exactly twice the maximum frequency in the modulating signal. The maximum frequency in the music or speech signal sent over AM broadcast radio is 5 kHz, and hence the bandwidth of an AM broadcast radio station is 10 kHz.

With frequency modulation, the bandwidth of the modulated carrier is at least twice the maximum frequency in the modulating signal, but can also be many times that. The maximum frequency in the music or speech signal sent over FM broadcast radio is 15 kHz, but the bandwidth of an FM broadcast radio station is 200 kHz. The additional bandwidth consumed by such forms of broadband frequency modulation guarantees immunity to the clicks, pops, and hisses that comprise noise. In fact, there is always a tradeoff between noise immunity and bandwidth, with increased immunity to noise requiring more bandwidth.

DIGITAL

The digital age is upon us, so proclaim many futurists. But why should we care? The sound quality from the audio compact disc certainly is superb, but the basic service of listening to recorded music has not been revolutionized or even changed much. The telephone signals carried across the country are digital, but here too the very nature of telephoning has not changed, and again we wonder, so what. What then is digital?

A waveform exists at each and every instant of time and can have any amplitude between its minimum and maximum range. Such waveforms are called analog. Analog signals can be represented in a different form in which this continuous variation is made discrete in both time and amplitude, and the amplitudes are represented by numeric digits. Such a representation is called digital. Sound waves, the typed characters of this book, and the variations in light captured by a TV camera can all be represented in a digital format.

Digital is exactly what the term implies: the use of digits or numbers to represent a signal or analog waveform. Imagine that I ask many people to estimate the length of my thumb by drawing a representation of it on a sheet of paper. The results will be many varying estimates. Suppose instead I state that the length of my thumb is 4.7 inches. Everyone will then write down the same 4.7 representing the 4.7-inch length of my thumb. Digits do not vary.

Another way of visualizing this is to imagine a glass partially filled with water. Suppose the glass is two-thirds full. Some people will think it is 62% full, others 71% full, and so forth. Everyone will have a slightly different estimate. This is a world of analog.

Suppose instead, we have only two options regarding our glass of water, namely, it is empty or it is full. We would then conclude that a glass about two-thirds full was full. Similarly, a glass about one-third full would be judged empty. The only time we would have some difficulty deciding would be if the glass were exactly half full, or half empty. Thus, anything less than half full is called empty, and anything more than half full is called full. This is the world of binary. Full is represented as a "1" and empty as a "0." In the world of binary, digits are either 1s or 0s. An individual binary digit is called a bit.

A sequence of bits can be used to encode some other quantity. In the example of our glass of water, we could use two bits in the combination 00 to represent an empty glass, 01 to represent a glass one-third full, 10 to represent a glass two-thirds full, and 11 to represent a full glass. In this case, any in-between fullness would have to be represented by the four available choices. Decimal numbers can also be represented by sequences of bits, as shown in the table, in what is called binary notation.

This use of bits in binary notation is like a representational code. Not only can water in a glass and decimal numbers be represented using bits, but text characters can also be encoded using bits. The American Standard Code for Information Interchange (ASCII) uses eight bits to encode the upper and lower case alphabet, decimal digits, and various symbols. In ASCII, the upper-case K is 01001011, and # is 00100011. Eight bits taken together is called a byte.

Decimal	Binary
0	000
1	001
2	010
3	011
4	100
5	101
6	110
7	111

ANALOG/DIGITAL CONVERSION

The process of converting an analog waveform to its digital representation begins with sampling the waveform every fraction of a second. Harry Nyquist, a researcher at Bell Labs, was the first to realize decades ago that a waveform could be sampled in time and then reconstructed perfectly by filling in the in-between potions of the waveform. Next, the amplitude variation of the waveform is quantized—a process in which only a finite number of values are allowed. Any amplitude falling in between is, in effect, made to be one of the allowed values. Finally, each of the finite number of allowed values is encoded as a unique combination of bits. For example, any amplitude falling between 0.5 and 0.6 volts might be encoded as 00101. The stream of digital bits is converted back to an analog waveform by performing the inverse of the process.

Analog signals are characterized by their bandwidths. Digital signals are characterized by their bit rates—the number of bits that must be sent per second. Analog signals occupy bandwidth. Digital signals also occupy bandwidth, since the variation of the digital waveform must be preserved. The bandwidth of a digital signal is at most half the bit rate but can be less with appropriate sophisticated modulation schemes.

The bit rate needed to send a digital signal can sometimes be reduced through compression. Many signals have much redundancy, and by encoding this redundancy appropriately, the bit rate can be reduced. Consider transmitting in a second a series of 100 bits in which each bit is a "1." The bit rate is 100 bits-per-second, or 100 bps. Rather than sending the 100 identical bits, a scheme could be used to state that the next 100 bits are all "1's" and this scheme might require only 20 bits. The bit rate would thus be reduced to only 20 bps to send the same information.

TV signals and speech signals have much redundancy and can be compressed. MPEG (for: Motion Picture Experts Group) is a compression standard currently being used, and developed further, for video. MPEG-1 achieves very good results at bit rates of 3 Mbps to 4 Mbps, although compression usually is a compromise quality. In the case of a compressed TV signal, fast changes from frame to frame might blur and some detail might be lost in each frame. In the case of speech, some naturalness might be lost. The table gives some signals and their bit rates in both uncompressed and compressed forms. One thing we notice is that television requires a bit rate about one-thousand times that for a telephone speech signal. The 1000-to-1 bandwidth differential we observed for analog holds also for digital.

One reason digital is so popular today is that each copy is identical to each other and to the original. Another reason is that digital transmission offers considerable immunity to noise. Rather than preserving the waveform against noise, digital involves simply deciding whether a zero or one has been sent which is a simple threshold decision. And a last reason digital is so popular today is the widespread availability of digital computers and digital integrated circuits.

SIGNAL	BASIC BIT RATE	COMPRESSED BIT RATE
text	8 bits per character	3 bits per character
television	80 million bps	• good quality 4 million bps • lower quality 1.5 million bps
telephone speech	64,000 bps	• good quality 12,000 bps • questionable quality 1200 bps

COMPUTERS

The digital computer is an electronic machine that manipulates data under the control of a stored set of instructions, called the computer program. The physical aspects of the computer itself are known as hardware, and the computer programs that control the machine are known as software. The data manipulated by the computer can be numbers, such as the numbers used in the computers that calculate our pay and our tax bills, or a representation of textual characters, such as the text typed by me into the computer that I used as word processor in writing this book. Computers are very flexible in that the same machine can be used for many different purposes depending upon the software used in the computer.

A computer has its own internal memory where both the program that controls the computer is stored and accessed and data that is being processed is stored temporarily. External storage of data and programs is on floppy disks that each contain about 1.5 mega bytes of information. CD-ROM is a form of optical memory that can store about 5 giga bytes of data using technology similar to the compact disc (CD) used for music. Floppy disks store data magnetically and can be erased and rewritten with new data. CD-ROM (for Compact Disc Read Only Memory) can only be read. New forms of optical storage that can be erased and rewritten has recently entered the marketplace.

Computers are programmable calculating machines, but these calculations and other manipulations of data are performed with astonishing speed. Computers do not think nor are they very good at the pattern recognizing and reasoning done by people. Although the computer HAL in Stanley Kubrick's movie *2001: A Space Odyssey* was able to recognize human speech and even read lips, in fact we are still very far from such computer abilities. Given all the computer power at use by the Weather Bureau, weather forecasting is still sometimes quite dramatically wrong.

Over three decades ago, I encountered my first computer during a summer job at an insurance company in Newark, New Jersey. A few years later, I was using a much larger computer at Bell Labs to process speech signals and also to generate computer art. The IBM 7090 computer that I used then at Bell Labs filled a large room, yet its power was much less than what I have today at my finger tips as I type this book into my Macintosh computer. Computer power as measured by processing speed and memory capacity has increased astonishingly over the years while simultaneously physical size and price have decreased. When I performed my investigations of the use of digital computers in the visual arts three decades ago at Bell Labs, no artist could afford the computers of those days, but today computers of far greater power are in many of our homes and are used by many artists.

TRANSMISSION MEDIA

Signals are carried over distance by being transmitted over a wide variety of transmission media. One medium that we are familiar with is copper wire. Copper wire connects the telephones in our home to the central office. This medium is a pair of conductors of copper wire insulated from each other and lightly twisted together— hence, the term twisted pair to describe them. Copper wire can be configured in the form of an inner conductor and an outer shield— hence, the term coaxial cable to describe this medium, shown in Fig. A4.

Over very short distances, twisted pair can carry much bandwidth. This should not be surprising since the ribbon wire that connects the indoor TV antenna to the TV set clearly can carry many TV signals. The problem is that the bandwidth, or signal carrying capacity, of copper wire in the form of a twisted pair depends strongly on distance. A mile of twisted pair can not carry much more than a few mega Hertz or a few mega bits-per-second. Coaxial cable has much more bandwidth over large distances. The coaxial cable used by CATV can carry as many as a hundred TV channels—a bandwidth of about 500 MHz—over the miles from the head end to your home. However, the signal carried over coaxial cable loses much of its amplitude along the way and must be amplified every few thousand feet.

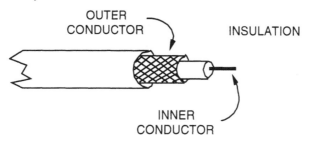

Fig. A4. In a coaxial cable, an inner conductor and an outer conductor form the electric circuit. The outer conductor shields the inner conductor from interference.

We are all familiar with radio signals. Radio carries over-the-air TV signals to our home and also AM and FM broadcast radio. Radio signals are electromagnetic waves that propagate through space and lose amplitude with distance.

Very-high-frequency radio signals have been used for long-distance telephone service. Such terrestrial radio operates in the microwave frequency bands and the radio waves travel mostly in a straight line of sight between towers located every 30 miles or so along the surface of the earth. Another approach is to send radio waves to a communication satellite located 22,300 miles above the earth's equator in what is called a geosynchronous orbit since the satellite is rotating around the earth in exactly 24 hours and thus appears stationary relative to the earth's surface.

A problem with geosynchronous satellites when used for two-way communication such as telephone service is the delay of about 1/4 second needed by the radio signal to travel to the satellite and back to earth. The round trip delay of the person at the distant end to respond is about 1/2 second, which impedes interactive communication and is not acceptable to most people. Communication satellites, however, are excellent for distributing one-way TV signals about the earth.

The newest transmission medium is optical fiber. Optical fiber carries light waves by continuously channeling the light waves along the axis of transparent, glass fiber that is only one-tenth the diameter of a human hair. Today's glass fibers are so pure that the amplitude of the light hardly diminishes over great distances. Although the amplitude of light can be made to vary in synchrony with the signal to be transmitted, the easiest approach is simply to turn the light on and off. Thus, most signals sent over optical fiber are encoded in a digital format.

The signal carrying capacity of optical fiber is nearly unlimited. Today's optical fiber systems easily convey a few billion bits-per-second. The theoretical capacity of fiber operating at a single frequency is about 200 billion bits per seconds. This means that a single optical fiber a tenth the diameter of a human hair could carry 3 million telephone conversations or as many as 4,000 TV channels. Given such a tremendous capacity, it is somewhat understandable that so much hype has been generated by optical fiber. The technological stomach to be filled is vast indeed and matched only by the greedy eyes of promoters!

NETWORKS

A CATV system consists of a head end where various TV programs are received and are then multiplexed together for transmission over coaxial cable to our homes. Each and every home receives the same multiplexed signal and thus the same programs, although the converter box at your TV set might be programmed so you can only decode the cable channels and programs you have payed for. The signal sent over the cable becomes weak with distance and must be amplified every few thousand feet along the way. The cable from your home is connected to the cable running outside your home by what is called a tap. The configuration of the cable network for a CATV system looks like a tree with a main trunk and various branches; it is called a tree network, as shown in Fig. A5.

The telephone system is quite different compared to CATV. For one difference, the telephone network is switched. For another difference, the bandwidth of a telephone signal is much less than the bandwidth of a TV channel. The telephone lines in your home are connected to the central location by a pair of wires that are lightly twisted together. Every home has its own pair of wires, and all the wire pairs are together in large cables. This type of network architecture is called a star network, as shown in Fig. A6.

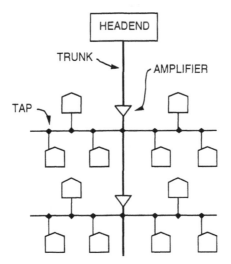

Fig. A5. A CATV system looks like a tree, with the coaxial cable network forming a trunk with branches to neighborhoods and homes. The signal on the cable becomes weak and must be amplified along the way.

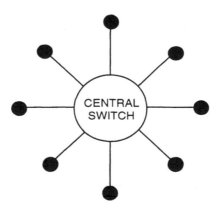

Fig. A6. The local telephone network is configured like a star. An individual pair of wires—called the local loop—connects each user to a central switch.

The central office is the place where all the wires terminate and encounter the first stage of switching. When you dial a telephone number, equipment at the central office determines what kind of circuits are needed to connect you to your called party and initiates the first stage of switching. If your call is long distance, the called number is transferred to equipment at your long distance company and switches make a connection for your call across the country. Many telephone calls are combined, or multiplexed, together on trunks and are carried over a variety of transmission media, as shown in Fig. A7. Time-division-multiplexing and optical fiber are the choice for most long-distance calls.

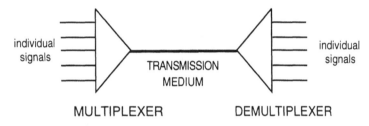

Fig. A7. With multiplexing, a number of individual signals are combined to share a transmission medium.

The switches used in the telephone network are sophisticated electronic machines that are controlled by digital computers. Switches of the past connected actual wires and paths together to create a connection, a technique called space-division switching, shown in Fig. A8. Today's switches route and connect digital bits for a fraction of a second, a technique called time-division switching. Either way, a complete circuit is created and maintained for the duration of your telephone call, a method called circuit switching.

Communication in the form of data between computers and in the form of text between people usually consists of short bursts. Maintaining a complete circuit connection for the full duration of a data call thus would not be efficient since the connection would be idle for most of the time. A more efficient technique is to break the data stream into short packets and to append the address of the destination on each packet. The packets of data can then be routed over the network and gradually make their way to the destination where the complete message might be reassembled. Such a form of switching is called packet switching. It is the form of switching used by the InterNet.

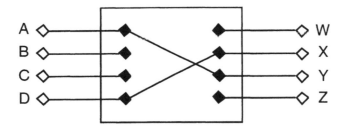

Fig. A8. With space-division switching, one circuit is physically connected to another. In this example, circuit A is connected to circuit Y, and D is connected to X.

READINGS

Noll, A. Michael , *Introduction to Telecommunication Electronics (Second Edition)*, Artech House: Norwood, MA, 1995.

Noll, A. Michael, *Introduction to Telephones & Telephone Systems (Second Edition)*, Artech House: Norwood, MA, 1991.

Pierce, John R. & A. Michael Noll, *Signals: The Science of Telecommunications,* Scientific American Library: New York, NY, 1990.

Appendix B

FINANCIAL TABLES

BABY BELL FINANCIALS
(1995)

Item	Ameritech Amount in millions	Margin	Bell Atlantic Amount in millions		Bell South Amount in millions		NYNEX Amount in millions		Pacific Telesis Amount in millions		SBC Corporation Amount in millions		US WEST, Inc. Amount in millions	
Operating Revenues	$13,428		$13,429		$17,886		$13,407		$9,042		$12,670		$11,746	
Operating Expenses	(7,948)		(7,716)		(10,057)		(8,748)		(5,167)		(7,463)		(6,810)	
CASH FLOW (EBITDA)	$5,480	41%	$5,713	42%	$7,829	44%	$4,659	35%	$3,875	43%	$5,207	41%	$4,936	42%
Depreciation & Amortization	(2,177)		(2,627)		(3,455)		(2,567)		(1,864)		(2,170)		(2,291)	
OPERATING INCOME	$3,303	25%	$3,086	23%	$4,374	24%	$2,092	16%	$2,011	22%	$3,037	24%	$2,645	22%
Interest	(469)		(561)		(724)		(734)		(442)		(515)		(527)	
Other Income & Expense	260		481		20		352		42		270		36	
PROFIT BEFORE TAXES	$3,094	23%	$3,006	22%	$3,670	20%	$1,710	13%	$1,611	18%	$2,792	22%	$2,154	18%
Taxes	(1,086)		(1,148)		(1,024)		(641)		(563)		(903)		(825)	
Accounting Changes & Extraordinary Item			(7)											
NET INCOME	$2,008	15%	$1,858	14%	$2,646	15%	$1,069	8%	$1,048	12%	$1,889	15%	$1,317	11%

Notes:
Bell South financials exclude restructuring charges of $1,082M & effect of change in accounting practices of $3,360M.
NYNEX financials exclude effect of change in accounting practices of $2,919M.
Pacific Telesis financials exclude effect of change in accounting practices of $3,360M.
SBC financials exclude effect of change in accounting practices of $2,819M.

FINANCIALS
(1995)

Item	AT&T (1994) Amount in millions	Margin	TCA Cable TV, Inc. Amount in millions		Tele-Communications, Inc. (TCI) Amount in millions		Time Warner Inc. Amount in millions		Time Warner Entertainment Company, L.P. Amount in millions		GTE Corporation Amount in millions		AirTouch Amount in millions	
Operating Revenues	$75,094		$189.1		$6,851		$8,067		$9,517		$19,957		$1,619	
Operating Expenses	(63,025)		(95.0)		(4,937)		(6,885)		(7,518)		(11,226)		(1,290)	
CASH FLOW (EBITDA)	$12,069	16%	$94.1	50%	$1,914	28%	$1,182	15%	$1,999	21%	$8,731	44%	$329	20%
Depreciation & Amortization	(4,039)		(28.3)		(1,372)		(559)		(1,039)		(3,675)		(216)	
OPERATING INCOME	$8,030	11%	$65.8	35%	$542	8%	$623	8%	$960	10%	$5,056	25%	$113	7%
Interest	(748)		(13.8)		(1,010)		(877)		(545)		(1,047)		22	
Other Income & Expense	236		0.3		177		214		(256)		(5)		245	
PROFIT BEFORE TAXES	$7,518	10%	$52.3	28%	($291)	-4%	($40)	-1%	$159	2%	$4,004	20%	$206	15%
Taxes	(2,808)		(21.0)		120		(126)		(86)		(1,466)		(113)	
Accounting & Other Items					(34)		(52)							
NET INCOME	$4,710	6%	$31.3	16%	($205)	-3%	($218)	-3%	$73	1%	2,538	13%	$132	8%

GLOSSARY

Amplifier. An electronic device that increases the power or amplitude of a signal. An amplifier makes a small signal large.

Analog. The representation of a signal in terms of its continuously varying shape.

Baby Bells. The affectionate name for the seven regional holding companies that own the local telephone companies that formerly were owned by AT&T.

Bandwidth. The width of the range of frequencies that describe a signal. Also, the width of the range of frequencies passed by a communication channel or device.

Bit. Stands for binary digit. The smallest entity for encoding digital information.

Bit Rate. The rate at which bits are transmitted. Usually expressed in bits per second, or bps.

Byte. Eight bits of data.

CATV. A system of transmitting television signals over coaxial cable. CATV originally stood for Community Antenna TeleVision—a system of using one tall master antenna to receive distant TV signals and then retransmit them over coaxial cable to a number of homes.

Circuit Switching. With circuit switching, individual connections are made and maintained for the duration of the call.

Coaxial Cable. A transmission medium consisting of two copper-wire electrical conductors. One conductor is along the axis and the other conductor surrounds it, separated by an insulator. This configuration isolates the conductors from external interference and offers much bandwidth, but the signal decreases greatly in amplitude with distance thereby requiring amplification.

DBS. Direct broadcast satellite television uses a communication satellite in geosynchronous orbit to broadcast television signals over a very wide areas, typically an entire continent.

Digital. The representation of a signal as a series of digits.

Frequency. The rate at which a signal varies or repeats itself, competing one full cycle. Frequency is measured in Hertz.

Frequency-Division Multiplexing. The combining together of a number of signals to share a transmission medium by assigning each signal its own unique range of frequencies.

Hertz (Hz). A measure of the frequency of a signal.

IXC. An IntereXchange Carrier is another name for a long-distance company.

Laser. An electronic component that creates light of a single frequency in which all the light rays are parallel to each other and are all synchronized in phase.

LATA. Local Access and Transport Area. The MFJ created a number of LATAs within the United States and restricted the Baby Bells to the provision solely of local service within these LATAs.

LEC. A Local Exchange Carrier is another name for a local telephone company.

MFJ. The Modified Final Judgement was signed by AT&T and the United States Department of Justice in 1982. The MFJ modified the Final Judgement of 1956 that prevented AT&T from entering the computer and other non-communication businesses. The MFJ resulted in the divestiture of the local telephone companies from AT&T that occurred on January 1, 1984..

Modem. A modulator/demodulator device used to interface a computer to the telephone network.

Modulation. The process of varying the maximum amplitude or instantaneous frequency of a carrier signal in direct proportion to the time variations of a modulating signal. The effect is to shift or move the frequencies of the modulating signal to a different frequency range.

MPEG (Motion Picture Experts Group). A system of standards for the compression of motion pictures and video. Compression reduces the bit rate required for the moving imagery and is based on redundancy within the images and also on redundancy from one image to the next.

Multiplexing. The combining together of a number of signals to share a transmission medium.

Optical Fiber. A transmission medium for carrying light—in effect, a light pipe. The fiber is made from high-purity silica. A single strand of optical fiber has tremendous bandwidth and thus can carry many multiplexed signals.

Packet Switching. With packet switching, data signals are broken into a series of short packets with each packet specifying its address. Individual packets are examined by switching machines along the way and are switched according to the availability of transmission capacity. Packet switching is particularly appropriate for the short, bursty nature of data traffic and enable many simultaneous users to share transmission capacity. Packet switching is used for the backbone network of the Internet.

Sampling. The process of representing the continuous time variation of a signal as a series of time intervals.

Spectrum. The depiction of the range of the various frequencies that comprise a signal, or of the range of the various frequencies passed by a communication channel or device.

Time-Division Multiplexing. The combining together of a number of signals to share a transmission medium by assigning the sample values of each signal their own unique intervals in time.

BIBLIOGRAPHY

Dutton, William H., Jay G. Blumler, & Kenneth L. Kramer (Editors), *Wired Cities: Shaping the Future of Communications*, G. K. Hall & Co. (Boston, MA), 1987.

> An excellent overview of the history of wired cities to the mid 1980s with many of the chapters written by the people responsible for the activities. Includes analyses of the policy issues associated with wired cities. The chapter "America's Talk-Back Television Experiment: QUBE" by Carol Davidge (pp. 75-101) is a detailed account of Warner's experience in two-way CATV.

Goldmark, Peter C., "Communications Technology for Urban Improvement," Report to the Department of Housing and Urban Development, Contract No. H-1221, Committee on Telecommunications, National Academy of Engineering, June 1971.

> This is the final report of the landmark study that instituted much of the work in the early 1970's on broadband networks that combine many services and also interactive, two-way capabilities.

Kay, Peg, "Social Services and Cable TV," The Cable Television Information Center, July 1976, NSF/RA-760161 (funded by the National Science Foundation).

> A thorough report of the beginnings of the wired nation along with detailed descriptions of all the possible uses of communication technology in delivering a wide variety of social services.

Mason, Edward S., *On the Cable: The Television of Abundance*, Report of the Sloan Commission on Cable Communications, McGraw-Hill Book Company (New York), 1971.

> Presents an understandable explanation of the workings of CATV from many perspectives, including technical and regulatory, and then advocates the continued deployment of CATV as a means of offering more program variety and expression.

Smith, Ralph Lee, *The Wired Nation*, Harper Colophon Books (New York), 1972.

> The vision of an electronic highway based on broadband communication over coaxial cable is clearly presented and forcibly argued. Although based on ideas nearly 25 years old, the vision is still crystal clear and well stated.

———

Listed below are many of the author's published academic papers that treat in more detail topics covered in this book.

> "Teletext and Videotex in North America: Service and System Implications," *Telecommunications Policy*, Vol. 4, No. 1 (March 1980), pp. 17-24.

> "Videotex: Anatomy of a Failure," *Information & Management*, Vol. 9, No. 2 (September 1985), pp. 99-109.

> "The broadbandwagon!," *Telecommunications Policy*, Vol. 13, No. 3 (September 1989), pp. 197-201.

> "Voice vs Data: An Estimate of Future Broadband Traffic," *IEEE Communications Magazine*, Vol. 29, No. 6 (June 1991), pp. 22, 24, 29, 78.

> "Anatomy of a failure: picturephone revisited," *Telecommunications Policy*, Vol. 16, No. 4 (May/June 1992), pp. 307-316.

———

Listed below are a selection of the author's op-ed pieces and articles that explore issues and questions raised in this book.

"Videophone: A Flop That Won't Die," *The New York Times*, Sunday, January 12, 1992, Sec. 3, p. 13.

"Monopoly Runs Amok: N.J. Bell Wins, the Public Loses," *Daily Record* (Morris County, NJ), January 1, 1993, p. A11.

"Baby Bells Should Stick With Strengths," *Los Angeles Times*, October 22, 1993, p. B15.

"The Phone Company Has Gone Hollywood," *Daily Record* (Morris County, NJ), January 7, 1994, p. A11.

"The Collapse of the 'Information Superhighway': A Perspective on the Failed Telco/CATV Deals," *Telecommunications*, Vol. 28, No. 7, July 1994, p. 19.

"Dialing for Dollars," *Los Angeles Times*, October 3, 1994, p. B7.

"Another Perspective on the Telco/CATV Debate," *Telecommunications*, Vol. 28, No. 12, December 1994, p. 20.

"Rethinking The Digital Mystique," *Telecommunications*, Vol. 30, No. 1, February 1996, p. 43.

"The Hazards of Cyber Overload," *Telecommunications*, Vol. 30, No. 3, March 1996, p. 44.

"Computers in the classroom: an idea that simply doesn't add up," *The Sunday Star-Ledger*, March 10, 1996, Sec. 10, p. 5.

"Aren't We Glad They Broke Up Bell?" *Los Angeles Times*, April 24, 1996, p. B13.

———

INDEX

About the Authors

A. MICHAEL NOLL is a professor at the Annenberg School for Communication at the University of Southern California and was dean there for a two-year interim period. He is an early pioneer in the use of digital computers in the visual arts and performed basic research at Bell Labs. He has worked in science policy at the White House and in marketing at AT&T. The author of six textbooks and many professional articles, Noll writes for various newspapers and a trade journal.

ABE M. ZAREM is chairman of his own consulting firm. He was founder, chairman, and CEO of Electro-Optical Systems, Inc. and Xerox Development Corporation.

Milton Keynes UK
Ingram Content Group UK Ltd.
UKHW022107141024
449569UK00031B/1805